An Introduction to
Mathematical
Taxonomy

G. DUNN
School of Epidemiology & Health Sciences
University of Manchester

B. S. EVERITT
Institute of Psychiatry
University of London

Dover Publications, Inc.
Mineola, New York

Next, in the potato, we have the scarcely innocent underground stem of one of a tribe set aside for evil; having the deadly nightshade for its queen, and including the henbane, the witch's mandrake, and the worst natural curse of modern civilization – tobacco ... Examine the purple and yellow bloom of the common hedge nightshade; you will find it constructed exactly like some of the forms of the cyclamen; and getting this clue, you will find at last the whole poisonous and terrible group to be – sisters of the primulas!

John Ruskin
The Queen of the Air

Bibliographical Note

This Dover edition, first published in 2004, is an unabridged republication of the work originally published in 1982 by Cambridge University Press as a volume in the series *Cambridge Studies in Mathematical Biology*.

International Standard Book Number: 0-486-43587-3

Manufactured in the United States of America
Dover Publications, Inc., 31 East 2nd Street, Mineola, N.Y. 11501

CONTENTS

PREFACE

Our aim in this text is to provide biologists and others with an introduction to those mathematical techniques useful in taxonomy. We hope that the text will be suitable both for undergraduates following courses in mathematical biology and for research workers whose interests include classification of particular organisms. The mathematical level demanded for understanding the text has been kept deliberately low, involving only a passing knowledge of matrix algebra and elementary statistics.

There are already a number of books available dealing with the topics of numerical and mathematical taxonomy, notably those of Sneath & Sokal, and Jardine & Sibson. This text, however, is specifically designed as an *introduction* to the area, and does not claim to be as comprehensive as those just mentioned. It should, however, serve as useful preparation for those two works.

Our thanks are due to Dr David Hand for many useful comments on the text, and to Mrs Bertha Lakey for her careful typing of the manuscript. Finally we would like to thank all of our colleagues at the Institute of Psychiatry for providing a stimulating working environment, and almost constant entertainment during the preparation of this book.

B. S. Everitt
G. Dunn

Institute of Psychiatry
London, November 1980

1

An introduction to the philosophy and aims of numerical taxonomy

1.1 Introduction

Classification of organisms has been a preoccupation of biologists since the very first biological investigations. Aristotle, for example, built up an elaborate system for classifying the species of the animal kingdom, which began by dividing animals into two main groups; those having red blood, corresponding roughly to our own vertebrates, and those lacking it, the invertebrates. He further subdivided these two groups according to the way in which the young are produced, whether alive, in eggs, as pupae and so on. Such classification has always been an essential component of man's knowledge of the living world. If nothing else, early man must have been able to realize that many individual objects (whether or not they would now be classified as 'living') shared certain properties such as being edible, or poisonous, or ferocious, and so on. A modern biologist might be tempted to remark that the earliest methods of classifying animals and plants by biologists were, like those of prehistoric man, based upon what may now be considered superficially similar features, and that although the resulting classifications could have been useful, for example for communication, they did not in general imply any 'natural' or 'real' affinity. However, one might then be tempted to ask what are 'superficially similar features' and what are 'real' or 'natural' classifications? Such questions and the attempts to answer them will be discussed in later parts of this chapter.

1.2 Systematics, classification and taxonomy

Before proceeding further it is necessary to introduce a number of terms which will be met frequently throughout the rest of the book. The definitions given here are intentionally brief; the full extent of the meaning of each term will become apparent during the remaining chapters.

Systematics – the scientific study of the kinds and diversity of organisms and of any and all relationships among them (Simpson, 1961).

Classification – the ordering of organisms into groups on the basis of their relationships. The relationships may be *genetic*, evolutionary (*phylogenetic*) or may simply refer to similarities of phenotype (*phenetic*).

Taxonomy – the theory and practice of classifying organisms (Mayr, 1969). (In the last two definitions it is important to distinguish classification, meaning the construction of classificatory systems, from the process of placing an individual into a given group, or the act of classifying, which is more properly referred to as identification; see Chapter 7.)

Once an ordering of organisms has been achieved one of course needs a means of referring to the classified groups; that is, one needs a convenient and informative method of *nomenclature*. The last word to be defined here is *taxon*. When one speaks of robins, lions, orchids or yeasts one is referring to the members of distinct groups of organisms, called taxa. A taxon is a taxonomic group of any rank that is sufficiently distinct to be worthy of being assigned to a definite category (Mayr, 1969). This definition implies that the delimitation of a taxon against other taxa of the same rank is virtually always subject to the judgement of the taxonomist.

1.3   The construction of taxonomic hierarchies by traditional and numerical taxonomy: comparison of methods

In order to summarize and make sense of the diversity of organisms the taxonomist customarily constructs a taxonomic hierarchy in which a taxon occupies a position in a nested scheme such as that given in Table 1.1, involving the classification of wolves, honeybees and common wasps. The hierarchy is intended to illustrate that different species within a given genus are more similar to one another than to species of other genera. Similarly, genera of one family are more similar to one another than they are to those of different families, and so on. Wolves are clearly not very similar to bees and wasps, but they are all classified as being members of the animal kingdom (Animalia), implying that they share some properties that are not characteristic of, say, members of the plant kingdom

Table 1.1. *A simple hierarchical classification*

Ranks	Taxa		
	Wolf	Honeybee	Wasp
Kingdom	Animalia	Animalia	Animalia
Phylum	Chordata	Arthropoda	Arthropoda
Class	Mammalia	Insecta	Insecta
Order	Carnivora	Hymenoptera	Hymenoptera
Family	Canidae	Apidae	Vespidae
Genus	*Canis*	*Apis*	*Vespula*
Species	*lupus*	*mellifera*	*vulgaris*

(Plantae). In addition, wasps and bees share properties not character-istic of wolves; that is, they are both Hymenoptera ('having mem-branous wings').

Taxonomy as a quantitative science is concerned with the problems of constructing such (usually) hierarchical structures, and in opera-tion consists of essentially four separate stages. First one has to decide on what one wishes to classify. On the assumption that one is able to distinguish living from inanimate material (by studying the history of science one can see that the distinction is by no means trivial), one could select, for example, deoxyribonucleic acid (DNA) sequences, proteins, organisms, species, or some more complex groups. In order to do this one has to have previous knowledge, or a previous system of classification, or else one would not be able to distinguish animate from inanimate objects, animals from plants, or daisies from orchids. No modern classification occurs in the absence of such previously formed classifications; one's knowledge is always built on previous experience, whether one ultimately rejects the previous ideas or merely adds to them.

Next one decides on the choice of characters on which to base comparisons between the taxonomic units (referred to as *operational taxonomic units*, OTUs, by numerical taxonomists). Now, despite the fact that numerical taxonomists sometimes claim that they choose as many characters as possible (Sokal & Sneath, 1963), this clearly cannot be true. Both the traditional taxonomist and the numerical taxonomist are forced to make subjective decisions on what sort of characters to select for comparison, but while there may, in practice, be differences in the way they choose these characters, the real difference between the two approaches lies in what they then do with the resulting observations; that is, in the assessment of similarity between units and in the use made of these similarities to construct the final classification.

The traditional taxonomist makes *intuitive* or *subjective* decisions concerning similarity, which, he claims, are based upon experience, skill and perhaps insight. The numerical taxonomist, on the other hand, bases his comparisons on an estimate of a defined measure of similarity (see Chapter 3), which is *objective* in the sense that the measure can be re-estimated by a second taxonomist using a different set of observations; such a procedure has a further advantage in being open to criticism in a way that an intuitive, subjective decision cannot be.

The final stage of the four is to make decisions concerning the classification of units on the basis of their previously assessed similarities. Again the traditional taxonomist will base such decisions on intuition, experience, and skill (he hopes!), whilst the numerical taxonomist resorts to a defined set of rules within one of the many *cluster analysis* techniques available (see Chapter 6). Which is the better method? *A priori* one cannot tell. However, there are situations where one can quite easily decide which is the easier, or more economical in terms of intellectual effort. For example, how does one effectively judge similarity between amino acid sequences of proteins without referring to a set of rules? Again, how does one assess a gradient or gradients of properties of characters across, for example, the British Isles, without resorting to some sort of defined quantitative measurements? It is in such situations and in many others that we feel that the methods of numerical taxonomy will be more applicable or more useful than the traditional approaches associated with the names of Linnaeus, Darwin or Mayr.

1.4   The philosophy of taxonomy

Consider a hypothetical situation in which one is asked to classify individuals within each of the following groups: warblers, hawkweeds, enteric bacteria, viruses, neolithic ceramics and rocks.

Are the methods of classifying rocks and ceramics applicable to the classification of living organisms? Are the methods of classifying warblers applicable to bacteria? Most biologists would answer 'no' to the first question, and many would give the same answer to the second. Why? Why do biologists often regard the classification of living material to be something special, needing its own particular logic or philosophy? It is not the purpose of this book to give final answers to these questions, but some discussion is needed since the numerical taxonomist explicitly denies that there are, or should be, any particular methodologies specifically applicable to the classification of the living world.

One does not have to read many textbooks on taxonomy to realize that there is no single underlying philosophy for this field, and one is tempted to conclude that 'anything goes' (Feyerabend, 1975). Much of the controversy appears to centre around the biologist's concept of a species. The typical view of a 'traditional' taxonomist (Simpson, 1961; Mayr, 1969) is that species (and often genera and higher taxa) are real entities that have to be discovered or revealed by the methods of classification. '... individuals do not belong in the same taxon because they are similar, but they are similar because they belong to the same taxon' (Simpson, 1961). The implication of such a view is that a particular classification is equivalent to a scientific theory, and so could be shown to be wrong. One particular difficulty of this belief is the necessity of producing a definition of species which is applicable to animals, plants and micro-organisms. Most of the traditional views concerning the definition of a species are irrelevant when one considers bacteria and viruses. Surely, even if one accepts the view that classification of organisms is, or should be, different from the classification of rocks, one needs to have a philosophy of taxonomy that will apply to all of the living world, and not just to, say, animals.

An alternative view is that

> Nature produces individuals and nothing more...species have no actual existence in nature. They are mental concepts and nothing more...species have been invented in order that we may refer to great numbers of individuals collectively. (Bessey, 1908)

Gilmour (1940) has summarized this alternative view of classification as follows:

> The classifier experiences a vast number of sense data which he clips together into classes...thus a class of blue things may be made for sense data exhibiting a certain range of colour, and so on...the important point to emphasize is that the construction of these classes is an activity of reason, and hence, provided they are based on experienced data, such classes can be manipulated at will to serve the purpose of the classifier...The classification of animals and plants...is essentially similar in principle to the classification of inanimate objects.

This is the philosophy of the numerical taxonomist. The implication of the numerical taxonomist's approach is that the resulting classification can be neither right nor wrong. It is not a theory, but merely a way of summarizing information in an intelligible form. One assesses its value by consideration of its usefulness to other biologists. If one accepts this view one can quite easily accept that traditional (evolutionary) and numerical (phenetic) taxonomies can exist side by side. One does not judge the classificatory method on the *a priori* beliefs of the taxonomist, but on the usefulness of the results, a view endorsed by Ruse (1973):

> A classification is a division based on a set of rules and, for this reason, is neither true nor false (which is what a theory is). This is not to deny that if, for example, evolutionary taxonomists can show that phenetic taxonomy is inferior to evolutionary taxonomy in its ability to enable taxonomists to summarize material or to predict things, then in this respect phenetic taxonomy is fair game. The proof of the pudding is in the eating, and if phenetic taxonomists cannot deliver what they claim to be able to deliver, then they are rightly open to criticism.

When assessing the utility of a particular approach to classification, one always has to bear in mind the reasons for which the classification was made. The first important role of any system of classification is as an aid to memory, particularly if the classification is hierarchic. Knowing where a particular taxon comes in a hierarchical scheme enables one to remember many of its characteristics (particularly if the characteristics are those which were originally used to construct the taxon concerned). The second role, very closely associated with the first, is as an aid to prediction of properties that have not been used to make the original classification. If, for example, one knows that orchids have a characteristic association with saprophytic fungi (a characteristic unlikely to have been used in the construction of the taxon Orchidaceae), it can be predicted with reasonable confidence that a plant identified as an orchid from its flower structure will also be growing in association with a fungus. Finally, an important function of any classification of the living world is its explanatory power, particularly with respect to the pathways of evolution. (This will be dealt with in greater detail in the next section.)

One argument for classifications produced by a numerical taxonomist, which fulfil these three roles at least as efficiently as those produced using traditional methods, lies in the amount of *information* utilized by each approach, the numerical taxonomist tending to use more, and more diverse, characters on which to base his classification. (This argument will be developed in the next chapter; see section 2.2.)

1.5 Classification and inferences concerning patterns of evolution

Virtually all present-day biologists believe in two fundamental concepts pertaining to the scientific study of the living world. The first is that of evolution through natural selection. The second is that of a universal genetic code; that is, the concept that all of the information required for the development of an organism is contained in coded sequences of nucleotide bases in deoxyribonucleic acid (DNA), or occasionally, as in some viruses, in ribonucleic acid (RNA). Evolution can be thought of as either the evolution of populations of organisms or of populations of nucleotide sequences, or both. Individuals clearly do not evolve in the above sense since they do not survive for more than a few years, at the most. What can the results of taxonomy tell one about the patterns of evolution?

Attempting to produce answers to this question is, intellectually, one of the most interesting uses to which a classification can be put, and it is here, perhaps, that one assesses the value of any particular method of classification.

It is vital that the student of evolution distinguishes *phenetic relationships*, which are based on the properties of organisms as they are observed now, from *phylogenetic relationships*, which describe the evolutionary pathways that have given rise to these organisms and their properties. The most important phylogenetic relationship is that expressed by a *genealogy*, and this is called a *cladistic relationship*. One can also define a *genomic relationship* between organisms based on the similarity of their DNA (or RNA) sequences. Now, one can use the methods of numerical taxonomy to classify organisms either on the basis of their phenetic relationships, or on the basis of their genomic relationships (or both). The latter can be obtained by the study of nucleotide sequences, or indirectly from the amino acid sequences of proteins. Finally, one hopes to infer phylogenetic or cladistic relationships from the resulting classification. It makes no difference to the argument whether the phylogenetic relationships are inferred from the classification itself or from the original distances or similarities; what is important is the fact that they are always inferred from phenetic or genomic relationships. Details of how this is done will be discussed later.

But how do the views of a zoologist such as Simpson differ from this? He claims that, since one knows that populations have evolved through a process of natural selection, one should assess one's system of classification from what is known about the past pathways of evolution, these being inferred from the study of, say, fossil evidence. The following statement taken from Simpson (1961) summarizes this point of view:

> It is preferable to consider evolutionary classification not as expressing phylogeny, not even as based on it (although in a sufficiently broad sense that is true), but as consistent with it. A consistent evolutionary classification is one whose implications, drawn according to stated criteria of such classification, do not contradict the classifier's view as to the phylogeny of the group.

Hence the term 'evolutionary taxonomist'. The difficulty of this approach is the problem of assessing phylogeny independently of a system of classification, and the argument has been rejected by numerical taxonomists as circular (Sokal & Sneath, 1963). How does one use fossil evidence to infer pathways of evolution without first classifying the fossils and in some way assessing their similarity to living organisms?

1.6 Summary

In this chapter we have discussed the general concepts of taxonomy and the different approaches that might be considered for constructing classificatory systems. But the reader might still ask why classify or, perhaps more realistically, why bother to reclassify when our interests should be directed towards much more 'exciting' fields such as molecular biology, developmental genetics or ecology? Why have biologists bothered to 'revise' the earlier classifications of Aristotle, Linnaeus or Darwin? We would answer that classification is an activity essential to all scientific work, and as the needs and knowledge of the scientist change so must the system of classification. Molecular biologists are just as dependent on an effective method of classification as was Mendel or Lamarck. Lack of knowledge of properties and groupings of organisms may have serious consequences for progress in particular areas, for example ecology and microbiology, where it has been estimated that perhaps half of the bacterial colonies on an agar plate inoculated with river water cannot be identified to species level, even by experts (see Sneath, 1978b).

The aim of this text is to introduce certain numerical or mathematical methods that have been used to help classify the biological world. The aim of this numerical approach is to rid taxonomy of its traditionally subjective nature and to provide *objective* and *stable* classifications. Objective in the sense that the analysis of the same set of organisms by the same sequence of numerical methods will produce the same classification; stable in that the classification remains the same under a wide variety of additions of organisms or of new characters. Whether these criteria are always met by the numerical methods will be the subject of discussion in later chapters, but we hope that the examples to be discussed in the remainder of the book will convince the reader that there are fields of taxonomy where numerical methods offer many advantages over the traditional approaches. If numerical methods enable the taxonomist to explore his data more easily, or if they can be used in situations where the traditional methods appear to be inapplicable or inefficient (for example in microbiology), why not use them?

2

Taxonomic characters

2.1   Introduction

Any taxonomic exercise begins with the choice of OTUs to be classified (usually organisms, but they could be, for example, populations or possibly even proteins), and a selection of characters, the states or forms of which are used to describe them. A *character* in this context may be defined to be any property that can vary between taxonomic units, and the possible values that it can be given are called the *states* of that character. Thus, for example, 'containing

Table 2.1. *Characters for classifying micro-organisms (Sneath, 1978b)*

Class of character	Examples
Morphological	Number of flagella
	Shape of spores
Physiological	Ability to grow anaerobically
Biochemical	Oxidase activity
	Acid production from galactose
Chemical constituents	Presence of lysine in the cell wall
Cultural	Usual appearance of colonies on a defined medium
Nutritional	Ability to grow on acetate as sole carbon source
	Requirement for thiamine
Drug sensitivities	Sensitivity to benzyl penicillin
Serological	Agglutination by an antiserum to a reference culture
	Presence of a specific precipitin band in a gel filtration experiment
Genetic	Percentage of GC in DNA
	Ability to be transduced by a given bacteriophage preparation
	Extent of pairing with a reference sample of DNA

a spore' is a character, with states 'yes' and 'no'. Similarly 'seed length' is a character, and '2.0 mm' is one of its states. As a further example, Table 2.1 shows the most commonly used classes of characters in the classification of micro-organisms (see Sneath, 1978*b*). The choice of OTUs is a reasonably straightforward step and usually not controversial; the choice of characters, however, is often subject to much controversy and involves a number of difficult problems. It is these problems that will be of central concern in the remainder of this chapter.

2.2   Number of characters

Any living organism, no matter how apparently simple, possesses a theoretically limitless number of characters which could be used to produce a classification. In practice, of course, one is limited in the number of characters that may be examined, simply because of temporal or economic considerations. It therefore becomes of practical importance to consider how many, and what type of characters to study. With micro-organisms, for example, it would be considered sensible to select characters from as many of the classes shown in Table 2.1 as possible. It is frequently suggested that at

Table 2.2. *List of characters and character states used in comparisons of populations of the red campion* (Silene dioica) (*Prentice*, 1980)

Character	Character states
1* Pedicel length	
2* Calyx length	
3 Calyx shape	Cylindrical/constricted-cylindrical/ conical/spherical/oval
4 Calyx nerves (anastomosis)	Anastomosing/not anastomosing
5 Red calyx pigment	Present/absent
6 Calyx glandular hairs	Absent or very sparse/present
7 Calyx hairs (straightness)	Straight/flexuous/crispate
8 Calyx hairs (stiffness)	Soft
9 Calyx-tooth shape	Acute/subacute/obtuse
10* Corolla diameter	
11 Corolla colour	18 colour-depth categories ranging from white to deep magenta
12 Petal dissection	Indented to less than half-way/indented to half-way or more
13 Additional petal lobes	Present/absent

14	Coronal scale colour	As petals, pink/not as petals, pink/not as petals, white
15*	Petal-claw length	
16*	Capsule length	
17	Capsule shape	Globose/ovoid/pyriform/long-pyriform
18	Capsule-tooth orientation	Erect/ascending/deflexed/curled back
19	Pedicel orientation[b]	Erect[a]
20*	Seed length	
21*	Seed length/breadth ratio	
22	Seed-back shape	Convex/flat/concave/rounded
23*	Seed-back width	
24	Seed-face type	Very convex/convex/flat/concave-convex/concave
25	Seed colour	From colour chart
26	Tubercle-tip colour	Black/dark brown/brown/ginger/chestnut/grey
27*	Seed-plate length	
28*	Seed-plate length/breadth ratio	
29*	Number of suture points per plate	
30*	Tubercle length	
31	Hylar-zone type	Prominent/level/recessed
32	Seed-surface granulation	Coarse/medium/fine/absent
33	Suture width	Very narrow/narrow/medium/wide
34	Suture outline	Sinuous/sharply-sinuous/serrate/lobate/stellate/digitate
35	Tubercle type	Prominent/level/recessed
36*	Plant height	
37	Stem glandular hairs	Absent or very sparse/scattered/dense
38	Stem clothing[c]	Shortly hairy/with long hairs
39	Stem-hairs (straightness)[c]	Straight/flexuous/crispate
40	Stem-hairs (orientation)[c]	Patent/deflexed
41	Stem-hairs (softness)[c]	Rather stiff/soft
42*	Number of internodes below inflorescence[d]	
43*	Length of lowest cauline leaf	
44	Shape of lowest cauline leaf	Lanceolate/ovate-acute/ovate-obtuse/rounded
45	Leaf glandular hairs (above)	Absent or very sparse/present
46	Leaf glandular hairs (below)	Absent or very sparse/present
47	Proportion of shoot with flowers	Less than half/half or more/nearly all

*Treated as quantitative.
[a] Invariant in present data set.
[b] When capsule ripe.
[c] On internode in mid-stem.
[d] From the ground to the lowest side-shoot bearing visible flower-buds.

least 50, but preferably 100 or more, characters should be used to produce a fairly stable and useful classification. However, such recommendations appear to be based on intuition, rather than on empirical evidence, and it might be possible in some circumstances to achieve a stable classification using fewer characters. Certainly the number of characters used in taxonomic studies varies widely, as can be seen in Tables 2.2 and 2.3. The first of these shows a set of characters that have been used to compare populations of red campion (*Silene dioica* L. Clairv.) by Prentice (1980). Table 2.3 shows a set of characters used to study variation of winged aphids (Jeffers, 1967). Possible drawbacks of selecting such a limited range of, essentially, morphological characters as those shown in Tables 2.2 and 2.3 will be discussed later; here it is sufficient to note that they correspond to only *one* of the categories of characters appearing in Table 2.1.

2.3   Type of characters and coding of character states

2.3.1   *Qualitative characters*
Qualitative characters may be simple two-state, presence or absence, features such as 4, 5, 6 and 13 of Table 2.2. For convenience one of these states is coded as 1 and the other as 0 (it does not

Table 2.3. *Characters selected for the study of variation of winged aphids* (*Jeffers*, 1967)

Character	Description	Character	Description
1	Body length	12	Leg length, tarsus III
2	Body width	13	Leg length, tibia III
3	Fore-wing length	14	Leg length, femur III
4	Hind-wing length	15	Rostrum $(+/-)$
5	Number of spiracles	16	Ovipositor $(+/-)$
6	Length of antennal segment I	17	Number of ovipositor
7	Length of antennal segment II		spines
8	Length of antennal segment III	18	Anal fold $(+/-)$
9	Length of antennal segment IV	19	Number of hind-wing
10	Length of antennal segment V		hooks
11	Number of antennal spines		

matter which). There may be multistate qualitative characters, such as coronal scale colour (character 14, Table 2.2) which is seen to have three possible states. Again, for convenience, these may be labelled 1, 2 and 3, but it should be remembered that the numbers have no quantitative significance, and therefore no arithmetical operations such as addition or multiplication, etc., ought to be performed on them. As an alternative to the labels 1, 2 and 3, one might wish to code the states of such a character in binary form, perhaps along the following lines:

Coronal scale colour	Binary characters		
As petals, pink	1	0	0
Not as petals, pink	0	1	0
Not as petals, white	0	0	1

Suppose such a coding scheme is adopted for this character. Should two plants coded by 100 now be regarded as having three character states the same, or only one? This problem will be taken up in the next chapter, but it is interesting to note that in the present example, if one did not distinguish between character matches on 1s and 0s, then the three-state character, 'coronal scale colour', would be given three times the 'weight' of a simple two-state character such as 'red calyx pigment' (character 5 of Table 2.2).

If, for some reason, data for a particular character are not available, or if one wishes to prevent comparison of particular states, a code is introduced to signify 'no comparison'. Here 'NC' will be used. As an example, consider the following binary coding for distinguishing bacterial colonies:

smooth colonies	1	0	0
rough colonies	0	1	0
mucoid colonies	0	0	1

and an alternative

smooth colonies	1	0	0
rough colonies	NC	1	0
mucoid colonies	NC	NC	1

On the assumption that two matching 0s are not recorded as indicating identical character states, and are ignored by whichever measure of similarity is to be used (see Chapter 3), two clones with the same colony morphology will be recorded as having one similar character state by each of these coding methods, and none that are different. However, clones that have different colony morphologies will be recorded as having no similar character states and two different ones by the first method of coding. On the other hand, the second method will indicate no similar character states, and only a single different one, and for this reason it might be preferred.

As a final example of the coding of qualitative character states consider the hypothetical data presented in Table 2.4, to be used to classify ten bacterial clones. How does one code these data sensibly? This problem is complicated by the fact that the spore and sporangial properties are only applicable if the bacterium sporulates. They are *secondary*, rather than *primary* characteristics. Should they therefore be given less 'weight'? Again, should the character 'colony texture' be given more or less weight than 'presence of spore'?

First consider 'colony texture' and 'presence of spore'. These will be coded initially as shown in Table 2.5. This seems to be quite reasonable as long as matching 0s are not counted as indicating

Table 2.4. *Phenotypic characteristics of ten hypothetical bacterial clones*

	Colony texture (rough/smooth/ mucoid)	Presence of spore (yes/no)	Spore shape (round/ oval)	Spore position (central/ terminal)	Sporangial shape (swollen/ normal)
Clone 1	Rough	No	NC	NC	NC
Clone 2	Smooth	No	NC	NC	NC
Clone 3	Smooth	Yes	Round	Central	Swollen
Clone 4	Rough	Yes	Oval	Terminal	Swollen
Clone 5	Smooth	Yes	Oval	Terminal	Normal
Clone 6	Rough	Yes	Round	Central	Swollen
Clone 7	Rough	Yes	Round	Central	Normal
Clone 8	Mucoid	Yes	Oval	Terminal	Normal
Clone 9	Smooth	Yes	Oval	Terminal	Swollen
Clone 10	Mucoid	No	NC	NC	NC

similarities. But what about the secondary characters? One might
be tempted to use the coding presented in Table 2.6. But one now
wishes to record matching 0s as similarities! One way round this
problem is to split each secondary character into two binary
characters. For example:

Spore shape	Binary characters	
round	1	0
oval	NC	1

Finally, one could, if required, weight the character 'presence of

Table 2.5. *Binary for colony texture and presence of spore*

	Colony texture			Presence of spore
Clone 1	1	0	0	0
Clone 2	NC	1	0	0
Clone 3	NC	1	0	1
Clone 4	1	0	0	1
Clone 5	NC	1	0	1
Clone 6	1	0	0	1
Clone 7	1	0	0	1
Clone 8	NC	NC	1	1
Clone 9	NC	1	0	1
Clone 10	NC	NC	1	0

Table 2.6. *Binary coding for morphological properties of spores and
sporangia*

	Spore shape	Spore position	Sporangial shape
Clone 1	NC	NC	NC
Clone 2	NC	NC	NC
Clone 3	1	1	1
Clone 4	0	0	1
Clone 5	0	0	0
Clone 6	1	1	1
Clone 7	1	1	0
Clone 8	0	0	0
Clone 9	0	0	1
Clone 10	NC	NC	NC

Table 2.7. *Binary coding for the phenotypic characteristics listed in Table* 2.4

	Colony texture			Presence of spore				Spore shape		Spore position		Sporangial shape	
Clone 1	1	0	0	0	0	0	0	NC	NC	NC	NC	NC	NC
Clone 2	NC	1	0	0	0	0	0	NC	NC	NC	NC	NC	NC
Clone 3	NC	1	0	1	1	1	1	1	0	1	0	1	0
Clone 4	1	0	0	1	1·	1	1	NC	1	NC	1	1	0
Clone 5	NC	1	0	1	1	1	.1	NC	1	NC	1	NC	1
Clone 6	1	0	0	1	1	1	1	1	0	1	0	1	0
Clone 7	1	0	0	1	1	1	1	1	0	1	0	NC	1
Clone 8	NC	NC	1	1	1	1	1	NC	1	NC	1	NC	1
Clone 9	NC	1	0	1	1	1	1	NC	1	NC	1	1	0
Clone 10	NC	NC	1	0	0	0	0	NC	NC	NC	NC	NC	NC

spore'. One simple, if crude, way of doing this is by listing the character more than once. It is listed four times in Table 2.7, which shows the complete coding for the ten clones. It is clear that there has now been *a priori* weighting of these characters. The coding scheme has been devised deliberately to give more weight to the character 'presence of spore'. This might be justified with respect to the secondary, sporulation-associated, characters, but is rather difficult to justify in comparison with 'colony texture'. Many microbiologists, however, might be quite happy with this type of weighting, even if they disagreed with the size of the actual weight given. Perhaps the most important point that this example illustrates, however, is how easy it might be for the unsuspecting taxonomist inadvertently to give too much weight to multistate or secondary characters. (This difficult question of character weighting will be returned to in section 2.4.)

2.3.2   *Quantitative characters*

A quantitative character is one that varies from one OTU to another in a way which may be counted or measured on an *interval scale*. Examples from Table 2.2 are characters 1, 2, 10, 15, 16, and so on. Most of those in Table 2.3 are quantitative. Such variables could be converted simply to binary characters. For example, consider measurements of mean bacterial cell length:

Cell lengths	Binary characters		
less than 0.5 μm	0	0	0
0.5–1.4 μm	0	0	1
1.5–3.0 μm	0	1	1
more than 3.0 μm	1	1	1

The number of columns used is one less than the number of increments decided upon. One will usually use a measure of similarity that includes matched 0s, and one should be aware that, as before, this method of coding implies that this character is receiving more weight than a simple two-state qualitative one. The increments need not be of equal size, so, for example, if one were recording level of resistance to drugs, exponentially increasing increments could be used.

However, such recoding of quantitative characters as binary means a great deal of lost information, and it is now considered preferable to deal with the quantitative measurements directly, although one might wish to transform the data in some way rather than use the raw measurements themselves. For example, the continuous measure of drug resistance, equivalent to the exponentially increasing increment mentioned above, would be the logarithm of the inhibitory drug concentration. In many cases one might wish to *standardize* quantitative measurements before using them to assess similarity, etc. One method is *ranging* (see Gower, 1971), where the raw measurement x is converted to x' as follows:

$$x' = (x - x_{min})/(x_{max} - x_{min}) \qquad (2.1)$$

where x_{min} and x_{max} are the minimum and maximum measurements obtained from the sample for the character in question. This allows a sensible comparison of organisms when characters differing widely in absolute value are being used. Fow example, it would give approximately equal 'weights' to measurements of body length and length of antennal segment I for the aphids whose characters are listed in Table 2.3. The reader should be aware that this is also a form of *a priori* weighting, as is the choice of units of measurement for these quantitative characters (on the assumption that they are not subsequently standardized).

A further form of standardization which might be very useful in particular situations is to adjust measurements, say length of limbs, etc., by dividing by a measure of overall size of the organism. After such a transformation most of the measurements will then be giving information concerning, essentially, the *shape* of the organism. A similar form of standardization for metabolic rates in micro-organisms could be made. Here the measurements are adjusted with respect to the growth rate of the culture under specified conditions so that the resulting measurements will then contain information concerning the *pattern* of metabolic processes for a particular clone.

This account should, we hope, have made the reader aware of the problems and potential pitfalls of even the relatively simple exercise of coding the data ready for subsequent analysis. Many of the decisions that have to be made are perhaps just as subjective as those made by a 'traditional' taxonomist. (The discussion of coding methods here is by no means exhaustive and readers interested in further aspects of the area are referred to Lockhart (1970) and Sneath & Sokal (1973).)

2.4   Weighting of characters

The issue of character weighting has already been introduced in an intuitive way in the previous section. Here we intend to extend the discussion of this, at times, controversial topic.

To weight a character means to give it greater or lesser importance than other characters when using these to produce a classification. There are several types of character weighting, some obviously necessary and consequently not controversial, others which cause much discussion and often disagreement amongst taxonomists. The first, often implicit, type of weighting occurs when one decides either to include or to reject a character from the analysis. Since there are, in theory, a virtually unlimited number of characters to choose from, one has to select a subset of these on which to base the classification, even when using a computer. The choice here is usually made on criteria such as ease of observation, availability of material, and experience. This type of weighting is called *selection* weighting. At the same time one may decide to reject other potential characters because they are logically correlated to previously selected characters (for example, the diameter of a plant stem when the circumference has already been selected) or because they appear to be invariant, and so on.

Apart from the inevitable procedures of selection or rejection there are two other forms of weighting – *a priori* and *a posteriori*. The former means that amongst the selected characters, some are considered more important than others; for example, it might be suggested that more reliance be placed on characters known to be good diagnostic features in other groups, or those assumed to be good indicators of phylogenetic relationships. This approach was criticized by Adanson in the eighteenth century, and again in the twentieth century by numerical taxonomists such as Sneath and Sokal, because it presupposes a knowledge of the classification one wants to produce before the analysis of the data. Most numerical taxonomists argue for an equal *a priori* weighting of characters, although a form of 'statistical weighting' of characters is sometimes considered acceptable (see Chapter 6). However, as was seen in the preceding section, the taxonomist using numerical methods needs to be aware that different approaches to coding character states, and the use of a particular similarity or distance measure (see Chapter 3), often imply subtle differences in character weighting. The real problem with *a priori* weighting is perhaps not that it is logically invalid, but that it is often very difficult to decide how to weight the characters in practice.

After a classification has been arrived at, one may wish to be able to *identify* new individuals using a diagnostic key or some form of discriminant function analysis (see Chapter 7). Here some characters will almost certainly be more useful than others, and so one uses *a posteriori* weighting of characters in the construction of such keys. This, however, is the last stage of the procedure, and should not influence the formation of the taxa concerned.

2.5 Homology of characters

When one makes a decision to compare the character states of different organisms or groups of organisms, one has to decide whether it is valid to compare a particular feature of organism A with a similar feature of organism B; that is, whether the two features or characters are *homologous*. Simpson (1961) has defined homology to be 'resemblance due to inheritance from a common ancestry'. One would not, for example, consider the leg of a cat and the leg of a beetle to be homologous, and the same would apply to the wing of a bat and the wing of a butterfly. One would, however, treat the arm of a man, the wing of a bird, and the front leg of a cat as homologous characters. Homology is often contrasted with *analogy*, resemblance due to common function. The wings of birds and butterflies are analogous since they are both used for flying, but are not homologous since they do not share any common ancestry.

The definition of homology given above has been the subject of considerable dispute. For example, Jardine & Sibson (1971) make the point that the definition is logically circular (since we can only infer phylogeny from the classification made after the comparison of characters) and that it also fails to provide any practical criterion for the determination of homologies. Sneath & Sokal (1973) are also critical of the usual definition of homology and prefer to use the term *operational homology* which implies resemblance between parts with respect to some set of properties. This leads to the idea of the classification of characters, or primary classification, as opposed to the classification of organisms, which may be called secondary classification (Jardine, 1967). One decides that two characters are homologous on the basis of resemblance measured according to some predetermined set of rules. An example used by Sneath & Sokal (1973) indicates how simple these rules might be in practice. Consider two species of insect that are both black, while others in the same genus are red. If one had no way of distinguishing the blackness of these two species, one considers them to have the same character state – black. Similarly, one would consider 'red' and 'black' to be states of the same character – body colour. If, however, it was found that the colouration of these insects could be due to both pigments and optical interference (such as iridescence), one would subdivide this character (body colour) into two: pigmental body colour and interference body colour.

The idea of a set of rules for indicating homology is well illustrated by the methods of comparing amino acid or nucleotide base sequences. Consider a hypothetical RNA fragment (where U, G, C and A represent uracil, guanine, cytosine and adenine, respectively):

(a) UCAGCAAUCCGU

and a second fragment:

(b) UCAGCAAUCCGUAAA

How does one decide whether these two sequences are homologous? One possible method is the following:

1. One considers the whole sequence to be a complex character, comprising several unit characters (the individual nucleotides) which can have states U, C, A or G.
2. One aligns the sequences in all possible ways. For example:
       UCAGCAAUCCGU
       UCAGCAAUCCGUAAA
       or
       UCAGCAAUCCGU
               UCAGCAAUCCGUAAA
       or
       UCAGCAAUCCGU
       UCAGCAAUCCGUAAA
   and so on.
3. For each possible alignment one counts the number of unit characters that match in the two aligned fragments. So, for the three alignments given above, the number of matches are one, three and twelve, respectively.
4. One selects the alignment that maximizes the match between sequences. If the number of matches is greater than a pre-assigned value one decides that the sequences are homologous.
5. Homologous unit characters (nucleotide bases) are those opposing each other during optimum alignment.

In practice, this particular algorithm might not be adequate, especially in situations where it is thought that there might have been, for example, internal additions or deletions of nucleotides during the evolution of the sequences. The reader interested in a more detailed discussion of this problem is referred to Fitch (1970).

2.6 Summary

The important and difficult issues of the choice of characters, the weighting of characters, and homology have been described only relatively briefly in this chapter. Much fuller discussions are available in Sneath & Sokal (1973) and Jardine & Sibson (1971). However, we will now assume that the organisms to be classified and the characters to describe them have been selected, and that numerical methods are required first to quantify their similarity, then to construct the classification, and finally to produce a diagnostic key; techniques for each of these procedures will be covered in the remaining chapters of this text.

3

The measurement of similarity

3.1 Introduction

Having recorded and coded the character states of the operational taxonomic units (OTUs) to be investigated, the taxonomist next proceeds to estimate their *similarity*; that is, their phenetic relationship, resemblance, or affinity. The measure of similarity used will depend on the types of character studied, and on the way in which the information has been coded. The complement of the similarity of two taxonomic units is their *dissimilarity* and in many cases it will be this measure that is determined from the data. The purpose of the present chapter is to introduce the commonly used measures of similarity or dissimilarity and indicate in which situations particular measures are applicable. (The term *proximity* will often be used in subsequent chapters to refer to both similarity and dissimilarity measures.)

3.2 Similarity measures for binary characters

Probably the simplest, and certainly the most common, type of character used by taxonomists is one that can occur as one of only two states, here designated as 1 and 0. As illustrated in Chapter 2, 1 might indicate the presence of a feature, and 0 its absence, or 1 and 0 may merely label two alternative forms of a character, such as 'red' or 'white'. If one considers any two OTUs, i and j, data for all of the character states recorded (assuming that they are all binary characters) can be summarized in a 2×2 table of counts having the following form:

	OTU i		
	1	0	
OTU j 1	a	b	$a + b$
0	c	d	$c + d$
	$a + c$	$b + d$	$p = a + b + c + d$

where p is the total number of binary characters studied, a is the number of characters where both OTUs have the code 1, b is the number of characters where OTU i is coded 0 and j 1, and so on. The number of matches is $a + d$, and the number of mismatches is $b + c$. This 2×2 table as used in numerical taxonomy is primarily a convenient arrangement of the data and should not be confused with conventional 2×2 tables which arise as the subject for a test of independence (see Sokal & Sneath, 1963, Chapter 6).

Many similarity coefficients have been proposed for binary data of this kind (see Sneath & Sokal, 1973; and Clifford & Stephenson, 1975, for extensive lists), but only two have been commonly used in numerical taxonomy, these being the simple matching coefficient and Jaccard's coefficient.

3.2.1   *The simple matching coefficient*

This coefficient appears to have been introduced into numerical taxonomy by Sokal & Michener (1958). It is defined as follows:

$$s_{ij} = \frac{a + d}{p} \qquad (3.1)$$

In words, it is the ratio of the total number of matches to the total number of characters. It ranges in value between zero, when the two OTUs fail to match on any of the p characters, to unity, when they match on every character. Since this coefficient involves only the total number of matches (whether they are 1s or 0s), it is particularly useful when it is considered that a match for 0s conveys the same amount of information as a match for 1s, as, for example, when the 1s and 0s are used merely as convenient labels for two alternative states of a character such as red and white. However, if a 1 is used to indicate the presence of some feature and 0 its absence, it *may* be necessary to consider alternative measures, such as Jaccard's coefficient, which exclude negative matches.

3.2.2   *Jaccard's coefficient*

This coefficient was introduced into taxonomy by Jaccard (1908). It differs from the simple matching coefficient by excluding the number of 'negative' matches but like the latter it ranges in value from zero to unity. The coefficient is defined as follows:

$$s_{ij} = \frac{a}{a+b+c} \tag{3.2}$$

In words, it is the ratio of the number of positive matches to the total number of characters minus the number of negative matches. The decision to include or exclude the number of matched 0s, that is d, is a difficult and, at times, a contentious one. In some situations it would be improper to neglect conjoint absences when estimating similarity, in others it would seem ridiculous to regard two taxa as similar largely on the basis of their both lacking some feature. (However, even if it were clearly justified to exclude some of the negative matches, it might be rather difficult to justify excluding them *all*; in part this can be overcome by the way the individual characters are coded by the taxonomist; see Chapter 2.) The following example from Sokal & Sneath (1963) illustrates some of the difficulties over the appropriate way to deal with negative matches when estimating similarity.

> The absence of wings, when observed among a group of distantly related organisms (such as a camel, louse and nematode), would surely be an absurd indication of affinity. Yet a positive character such as the presence of wings (or flying organs defined without qualifications as to kind of wing) could mislead equally when considered for a similarly heterogeneous assemblage (for example, bat, heron and dragonfly). Neither can we argue that absence of a character may be due to a multitude of causes and that matched absence in a pair of OTUs is therefore not 'true resemblance', for, after all, we know little more about the origins of matched positive characters.

Such comments imply that each particular application must be considered on its merits and that no absolute statement can be made on whether or not to include 'negative' matches, although Sokal & Sneath (1963) suggest that a reasonable and logically defensible position appears to be the inclusion of positive *and* negative matches for those characters which vary within the group under study.

Table 3.1. *Hypothetical data set for four OTUs on five binary characters*

		Character				
		1	2	3	4	5
	1	1	1	0	0	1
OTU	2	1	0	0	0	0
	3	1	1	1	0	1
	4	0	0	0	1	1

To assist readers to familiarize themselves with these coefficients generally and to illustrate differences between the simple matching coefficient and Jaccard's coefficient, the two coefficients will be calculated for the data shown in Table 3.1 which represents four OTUs scored on each of five binary characters. By calculating a particular coefficient for each pair of OTUs shown in this table we arrive at the *similarity matrix* for the set of OTUs. Each entry in such a matrix shows the similarity between one pair of OTUs. The similarity matrices for the simple matching coefficient and for Jaccard's coefficient calculated on these data are given in Table 3.2. (Two points to note about these similarity matrices are, first, that self-similarities are unity, and secondly, that the matrices are *symmetric* since the similarity between OTUs *i* and *j* is the same as the similarity between *j* and *i*).

Examination of the entries in the two matrices indicates that the two similarity coefficients can take quite different values for the same

Table 3.2. *Similarity coefficients obtained from data in Table* 3.1

	Simple matching coefficient similarity matrix					Jaccard's coefficient similarity matrix			
OTU	1	2	3	4	*OTU*	1	2	3	4
1	1.0				1	1.0			
2	0.6	1.0			2	0.33	1.0		
3	0.8	0.4	1.0		3	0.75	0.25	1.0	
4	0.4	0.4	0.2	1.0	4	0.25	0.0	0.2	1.0

pair of OTUs. This would be relatively unimportant if the coefficients were jointly *monotonic*, in the sense that if all OTU pairs for one of the coefficients were ordered so that their similarity values formed a monotonic series (that is, a series that either increases or decreases over the whole of its length), then those for the other coefficient also formed such a series. (Such joint monotonicity is a useful property since it would ensure that identical classifications of the OTUs would be obtained from either coefficient when certain clustering methods were used; see Chapter 6.) However, joint monotonicity is the exception rather than the rule for similarity coefficients and this can be illustrated with the similarities calculated above. Ordering the simple matching coefficient values to form a monotonically increasing series we have:

OTU pair	3,4	2,4	1,4	2,3	1,2	1,3
Simple matching coefficient	0.2	0.4	0.4	0.4	0.6	0.8

but taking Jaccard's coefficient for the same pairs gives:

Jaccard's coefficient	0.2	0.0	0.25	0.25	0.33	0.75

which is *not* monotonically increasing. The lack of this property for many similarity coefficients again implies that careful consideration must be given to the choice of coefficient to be used in any application, since different coefficients will in many cases lead to different final classifications. (Sneath & Sokal (1973) do, however, indicate the possible argument that different coefficients *should* lead to different taxonomic structures, for it can be shown that different coefficients estimate different aspects of the taxonomic relationship.)

3.3   Similarity measures for qualitative characters having more than two states

The measurement of similarity for data involving qualitative characters with more than two states can be handled by extending in an obvious way the coefficients for binary characters described in the previous section. For example, suppose two OTUs are scored on five characters each of which can take on four states *A*, *B*, *C*, *D*, with the following results:

	Character				
	1	2	3	4	5
OTU 1	A	B	C	D	C
OTU 2	A	C	D	D	B

The simple matching coefficient for these two OTUs would be given, as before, by (number of matches)/(number of characters), so taking the value 0.4.

However, with this approach no allowance is made for the possibility that a match in a four-state character is less likely than in a two-state character. A more elaborate treatment of qualitative characters with more than two states which considers the probability of a given match taking place is given by Smirnov (1960, 1969). He suggests that the similarity based upon any one character is weighted as a function of the probability of the simultaneous occurrence of such a character state in two separate OTUs; if these OTUs share a rare character state, this is given large weight; if, however, they share a commonly occurring state, then this is given lower weight. Although this idea of weighting characters on the basis of their rareness of occurrence has a certain attraction, it has been criticised on various grounds by Sneath & Sokal (1973) and they do not recommend Smirnov's coefficient for use in numerical taxonomy. An alternative, and perhaps more straightforward, way of solving this problem is to code the data as several binary characters (see Chapter 2).

3.4 Similarity measures for quantitative characters

Quantitative characters such as length or diameter could be dealt with by simply converting them to binary characters. For example, one could take length 'less than 2 cm' as one state, and length 'greater than 2 cm', as the other. Such a procedure obviously entails a loss of information and one might perhaps prefer to consider similarity measures not requiring this conversion. One such measure which has been widely used is equivalent in form to Pearson's product moment correlation coefficient; however, its use in the context of numerical taxonomy is far more contentious than its relatively non-controversial role in assessing the linear relationship between pairs of variables such as height and weight. When used as a measure of similarity for two OTUs its calculation involves averaging over the states of different quantitative characters to produce an 'average character state' for each OTU. Jardine & Sibson (1971) remark that such a procedure is 'absurd'.

It has often been suggested that the correlation coefficient will be a useful measure of similarity in those situations where absolute size alone is regarded as less important than shape. Thus, in classifying plants and animals, the absolute sizes of the organisms or their parts are often of less importance than their shapes, and in such cases the investigator requires a similarity coefficient which will take the value unity whenever the set of character-state values describing two OTUs are parallel, irrespective of how far apart they are; for example, the following sets of scores are parallel in this sense:

OTU 1	10	5	15	3	20
OTU 2	15	10	20	8	25
OTU 3	30	25	35	23	40

The correlation coefficient meets this requirement (it takes the value unity for each pair of OTUs in the above example); unfortunately, the converse is not true since the correlation may take the value one even when the two sets of values are *not* parallel. All that is required for perfect correlation is that one set of scores be linearly related to the second set (see Fleiss & Zubin, 1969, for an example).

Because of these points, and others made by Eades (1965) and Minkoff (1965), the correlation coefficient used as a measure of similarity must be regarded as unsatisfactory.

3.5 Measures of dissimilarity and distance

As mentioned in the introduction of this chapter, similarity and dissimilarity are mutually dependent terms. Given any similarity measure, s, the corresponding dissimilarity might simply be defined as $1 - s$, so that identical OTUs would have similarity unity and dissimilarity zero. A number of dissimilarity measures arise, however, that have no direct counterpart in the similarity measures discussed in the previous sections, and these will be described here. A number of these satisfy a set of mathematical properties which make them particularly attractive. Designating the dissimilarity between OTU i and OTU j as d_{ij}, these properties are as follows:

(a) *Symmetry*:

$$d_{ij} = d_{ji} \geq 0 \qquad (3.3)$$

That is, the dissimilarity between i and j is independent of the direction in which it is measured, and must be positive provided the two OTUs are not coincident.

(b) *Distinguishability of non-identicals*:

if $d_{ij} \neq 0$ then $i \neq j$ \qquad (3.4)

That is, d_{ij} will take some non-zero value if i and j are not the same OTU.

(c) *Indistinguishability of identicals*: Given two identical OTUs i and j, the dissimilarity d_{ij} is zero; in particular

$$d_{ii} = 0 \qquad (3.5)$$

(d) *Triangular inequality*: Given three OTUs i, j and k, the dissimilarities between them satisfy the inequality

$$d_{ik} \leq d_{ij} + d_{jk} \qquad (3.6)$$

The triangular inequality is also known as the metric inequality, and dissimilarity coefficients satisfying the above properties are known as metrics and generally referred to as *distances* rather than dissimilarities.

The most familiar metric is, of course, *Euclidean*; it is familiar because we live in a locally Euclidean universe and this tends to give it advantages in numerical taxonomy, where it is very widely used, because our daily experience gives us an intuitive grasp of Euclidean distances and thereby enables us to grasp their properties without difficulty. For a simple, two-character example, consider Figure 3.1(a). From Pythagoras' theorem it is clear that the Euclidean distance between OTU i and OTU j is given by the following expression:

$$d_{ij} = \left[(x_{ja} - x_{ia})^2 + (x_{jb} - x_{ib})^2 \right]^{\frac{1}{2}} \tag{3.7}$$

This expression is easily generalised to the situation where there are p characters observed on each OTU:

$$d_{ij} = \left[\sum_{k=1}^{p} (x_{ik} - x_{jk})^2 \right]^{\frac{1}{2}} \tag{3.8}$$

Such an expression arises from assuming that the p characters are represented by p *orthogonal* axes (i.e. axes at right angles to each other). However, in practice, because of the correlation of character states, this assumption will not be justified and the Euclidean distance will be a poor measure of the actual distance between OTUs i and j. (See Figure 3.1(*b*) for a two-dimensional example.) This problem can be overcome either by using oblique coordinate axes and a measure such as *Mahalanobis' generalised distance* (see later) or by transforming to *principal component* axes (see Chapter 4).

Since d_{ij} increases with the number of characters some authors recommend computing an average distance obtained from

$$\bar{d}_{ij} = \sqrt{(d_{ij}^2/p)} \tag{3.9}$$

For binary characters this average Euclidean distance is simply $(1 - s)^{\frac{1}{2}}$ where s is the simple matching coefficient described in section 3.2. For quantitative characters, Euclidean distance calculated directly on the raw data may make little sense where the characters have different scales. For example, suppose that three OTUs A, B and C have each been measured on the two characters, weight (in pounds) and height (in feet), with the following results:

	Weight (*lb*)	Height (*ft*)
A	60	3.0
B	65	3.5
C	63	4.0

The squared Euclidean distances are as follows:

$$d_{AB}^2 = (60 - 65)^2 + (3.0 - 3.5)^2 = 25.25$$
$$d_{AC}^2 = (60 - 63)^2 + (3.0 - 4.0)^2 = 10.00$$
$$d_{BC}^2 = (65 - 63)^2 + (3.5 - 4.0)^2 = 4.25$$

However, if height had been measured in inches these squared distances become

$$d^2_{AB} = 61 \qquad d^2_{AC} = 153 \qquad d^2_{BC} = 40$$

OTU A is now closer to OTU B than to C. Therefore, even when all the characters are uniquely determined except for scale changes, Euclidean distance will not even preserve distance rankings. Because of this, and despite other problems with such transformations (see Fleiss & Zubin, 1969), characters should be *standardized* before calculating Euclidean distances. By this is meant expressing each state in standard deviation units. For example, in the data used

Figure 3.1. Examples of (*a*) the appropriate use of Euclidean distance, i.e. orthogonal axes, and (*b*) the inappropriate use of Euclidean distance, i.e. non-orthogonal axes.

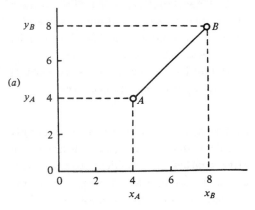

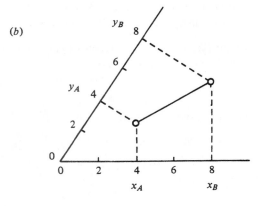

above the Euclidean distances would be calculated not from the raw
data but from the values

	Standardized weight	Standardized height
A	23.81	6.0
B	25.79	7.0
C	25.00	8.0

calculated by dividing each column of values by the standard
deviation of the three character values.

Penrose (1954) has suggested dividing the Euclidean distance
coefficient into two parts, a coefficient of 'size' and a coefficient of
'shape', using the following relationship

$$d_{ij}^2 = (p-1)C_Z^2 + pC_Q^2 \tag{3.10}$$

where d_{ij}^2 is the square of the Euclidean distance between OTUs i and
j, C_Z^2 is the proposed 'shape' coefficient given by

$$C_Z^2 = \frac{1}{p-1} \sum_{k=1}^{p} (x_{ik} - x_{jk})^2 - \frac{1}{p(p-1)} \left[\sum_{k=1}^{p} (x_{ik} - x_{jk}) \right]^2 \tag{3.11}$$

and the proposed size coefficient C_Q^2 is simply

$$C_Q^2 = \frac{1}{p^2} \left[\sum_{k=1}^{p} (x_{ik} - x_{jk}) \right]^2 \tag{3.12}$$

The shape coefficient, C_Z^2, looks formidable but is essentially nothing
more than the variance of the differences in the character states of
the OTUs being compared. It is likely to be large when considerable
discrepancy in the magnitude of the differences occurs, including a
mixture of positive, negative and negligible terms.

The size coefficient, C_Q^2, will be large when the character states
of the two OTUs are quite different in magnitude, and the differences
are largely in one direction. For example, a large C_Q^2 value would
arise if one OTU was very similar to another but much larger along
most of the character scales. (Sneath (1968) has introduced a similar
coefficient to allow for differing growth rates of microbial cultures
during the measurement of metabolic rates.)

In many studies in numerical taxonomy this partition into size
and shape coefficients may not be of great importance, but it may

be useful in a minority of investigations which involve comparing organisms of widely different sizes. However, it should be mentioned that in such situations Rohlf & Sokal (1965) have found Pearson's product moment correlation coefficient (see section 3.4) a more satisfactory measure of the similarity in shape of two OTUs.

Although Euclidean distance has been the dissimilarity measure most widely used in numerical taxonomy, as we shall illustrate by examples given in later chapters, a number of other measures have been employed in particular applications. One such measure is the *absolute* or *city-block* metric given by

$$d_{ij} = \sum_{k=1}^{p} |x_{ik} - x_{jk}| \qquad (3.13)$$

(When divided by the number of characters this measure is also known as the *mean character difference*; see Cain & Harrison (1958).) This has been used in anthropology by Czekanowski (1909, 1932), and by Haltenorth (1937) in a study of eight species of the large cats. Carmichael & Sneath (1969) also prefer this coefficient to Euclidean distance in their TAXMAP clustering procedure.

The rationale given by these authors for the use of this metric is that when, for example, two OTUs are specified by two characters whose scale units are of equal value they should be regarded as being the same distance apart whether (*a*) they are two units apart on each variable or (*b*) they are one unit apart on one variable and three units apart on the other. The use of the city-block metric satisfies this requirement, although one might argue that the requirement in many situations is not particularly convincing (but see below). The city-block metric certainly has the advantage of simplicity when compared with the Euclidean distance, although this is of little consequence when computers are used to calculate the dissimilarity or distance matrix. It has disadvantages when compared with Euclidean distance, the main one being its lack of invariance under rotations of the character space. This has serious implications for the interpretation of the results of ordination, as will be seen in the next chapter.

For an example of a situation where the city-block metric is the obvious measure of distance to use, consider the comparison of amino acid sequences in homologous proteins. Given below are the sequences of the last eight amino acids of cytochrome c from man, dog, and chicken:

Man	Tyr	Leu	Lys	Lys	Ala	Thr	Asn	Glu
Dog	Tyr	Leu	Lys	Lys	Ala	Thr	Lys	Glu
Chicken	Tyr	Leu	Lys	Asp	Ala	Thr	Ser	Lys

One way of measuring the distance between any two sequences is simply to count the number of non-matching amino acids. The distances calculated in this way for these three sequences are given in Table 3.4(*a*). Remembering that each amino acid is coded by a nucleotide triplet, one can calculate a *mutation distance* between two sequences, defined as the minimal number of nucleotides that would need to be altered in order for the gene for one sequence to code for the other (Fitch & Margoliash, 1967). The distances for the above

Table 3.3. *Mutation values for amino acid pairs*

	A	C	E	F	G	H	I	L	M	N	O	P	Q	R	S	T	U	V	W	Y
Aspartic acid (A)	0	2	2	2	1	1	2	1	3	1	1	2	2	2	2	3	2	1	2	1
Cysteine (C)		0	2	1	3	2	3	2	3	2	1	2	3	1	1	1	2	2	2	1
Threonine (E)			0	2	2	2	1	1	1	1	2	1	2	1	1	2	2	2	1	2
Phenylalanine (F)				0	3	2	3	2	2	2	1	2	3	2	1	2	1	1	1	2
Glutamic acid (G)					0	2	1	1	2	2	2	2	1	2	2	2	2	1	3	1
Histidine (H)						0	2	2	3	1	1	1	1	1	2	3	1	2	2	2
Lysine (I)							0	2	1	1	2	2	1	2	2	2	2	2	2	2
Alanine (L)								0	2	2	2	1	2	2	1	2	2	1	2	1
Methionine (M)									0	2	3	2	2	1	2	2	1	1	1	2
Asparagine (N)										0	1	2	2	2	1	3	2	2	1	2
Tyrosine (O)											0	2	2	2	1	2	2	2	2	2
Proline (P)												0	1	1	1	2	1	2	2	2
Glutamine (Q)													0	1	2	2	1	2	3	2
Arginine (R)														0	1	1	1	2	2	1
Serine (S)															0	1	1	2	1	1
Tryptophan (T)																0	1	2	3	1
Leucine (U)																	0	1	1	2
Valine (V)																		0	1	1
Isoleucine (W)																			0	2
Glycine (Y)																				0

Each value is the number of nucleotides that would need to be changed in order to convert a codon for one amino acid into a codon for another. The table is symmetrical about the diagonal of zeros. Letters across the top represent the amino acids in the same order as in the first column. From Fitch & Margoliash (1967).

Table 3.4. *Distance measurements between man, dog and chicken, based on the sequence of the last eight amino acids of cytochrome* c

(a) City-block metric, based on distances of unity between differing homologous amino acids

	Man	Dog	Chicken
Man	0	1	3
Dog	1	0	3
Chicken	3	3	0

(b) City-block matric, based on minimum mutation values (see Table 3.3) between differing homologous amino acids

	Man	Dog	Chicken
Man	0	1	4
Dog	1	0	5
Chicken	4	5	0

(c) Euclidean distance, based on the simple matching coefficient

	Man	Dog	Chicken
Man	0	$\sqrt{\frac{1}{8}}$	$\sqrt{\frac{3}{8}}$
Dog	$\sqrt{\frac{1}{8}}$	0	$\sqrt{\frac{3}{8}}$
Chicken	$\sqrt{\frac{3}{8}}$	$\sqrt{\frac{3}{8}}$	0

Note that (c) is jointly monotonic with (a), and that it would be with (b) if one were comparing nucleotide, rather than amino acid, sequences.

three sequences are given in Table 3.4(b). For each pair of amino acids a *mutation value* is taken from Table 3.3 which gives the minimum number of nucleotide changes required to convert the coding from one amino acid to the other. So, for example, if one wishes to determine the mutation value for glutamic acid and lysine, one notes that the codons for glutamic acid are GAA and GAG and those for lysine are AAA and AAG. The minimum number of nucleotide changes required to convert the codon from one to the other is 1; that is, one can either change from GAA to AAA or from GAG to AAG (or vice versa). If one were comparing nucleotide sequences in DNA, the mutation distance between any two sequences would be the number of non-matching nucleotides. If one considers each amino acid (or nucleotide) position as a separate character, it is not difficult to see why the above measurements of distance are logically equivalent to the city block metric. It is also intuitively satisfying to equate a mutation value of, say, 2 at one position with values of 1 at two other positions.

Further distance measures might be derived from the similarity measures described in section 3.2. The simple transformation, $1 - s$, has already been mentioned, but there are others that would be considered more attractive because of their metric properties. Gower (1966), for example, has shown that $(1 - s_m)^{\frac{1}{2}}$ and $(1 - s_j)^{\frac{1}{2}}$, where s_m is the simple matching coefficient and s_j is Jaccard's coefficient, are both Euclidean distances. See Table 3.4(c) for the distances between the above amino acid sequences, calculated using the simple matching coefficient.

3.6 Gower's similarity coefficient

In many applications of numerical taxonomy the OTUs will be described by a mixture of binary characters, qualitative characters with more than two states, and quantitative characters. In such cases a particularly useful measure of similarity is that proposed by Gower (1971) which is defined as follows:

$$s_{ij} = \sum_{k=1}^{p} s_{ijk} \bigg/ \sum_{k=1}^{p} w_{ijk} \qquad (3.14)$$

The weight w_{ijk} is set equal to 1 or 0 depending on whether the comparison of OTUs i and j is considered valid for character k and, except for the case of dichotomous characters, this weight can only be zero when character k is unknown for one or both OTUs. With dichotomous variables w_{ijk} is also set to zero when character k is known to be absent from both OTUs. Whenever $w_{ijk} = 0$, then s_{ijk} is set equal to zero, and if $w_{ijk} = 0$ for all characters, s_{ij} is undefined.

(a) *Binary or dichotomous characters.* The scores and weights for this type of data are as follows:

Individual i	+	+	−	−
Individual j	+	−	+	−
Score s_{ijk}	1	0	0	0
Weight w_{ijk}	1	1	1	0

If all characters were of this type, Gower's coefficient is simply Jaccard's coefficient as described in section 3.2.

(b) *Qualitative characters with more than two states.* In this case $s_{ijk} = 1$ if the two individuals i and j are the same for the kth character and zero if they differ. The number of states in the character is not taken into account, and for this type of data Gower's coefficient is equivalent to the simple matching coefficient discussed earlier.

(c) *Quantitative data.* In this case

$$s_{ijk} = 1 - |x_{ik} - x_{jk}|/R_k \qquad (3.15)$$

where x_{ik} is the value for OTU i on character k, and R_k is the range of the character values obtained either from a knowledge of the total range of the character in the relevant population, or as observed in the sample of OTUs under investigation. Here s_{ijk} will equal one when the character states of the two OTUs are identical and will be zero when these span the extremes of the range of character k. If all characters were of this type Gower's coefficient would essentially be equivalent to the complement of the absolute distance measure discussed in the previous section.

To illustrate the calculation of Gower's similarity coefficient, data from four hypothetical bacterial clones (See Table 3.5) will be analysed. First consider the similarity between clones 1 and 2. Taking the characters in the order that they appear in the table:

$$s_{12} = \frac{1 + (1 - 0.3/0.5) + (1 - 13/40) + 0 + 0}{1 + 1 + 1 + 1 + 1}$$

$$= 0.41$$

Table 3.5. *Phenotypic properties of four hypothetical bacterial clones*

	Presence of spores	Mean cell diameter (μm)	GC in DNA (%)	Colony morphology	Colony colour
Clone 1	+	1.0	57	smooth	brown
Clone 2	+	0.7	70	rough	white
Clone 3	−	0.5	30	mucoid	yellow
Clone 4	−	0.7	40	smooth	yellow

Now consider clones 3 and 4:

$$s_{34} = \frac{0 + (1 - 0.2/0.5) + (1 - 10/40) + 0 + 1}{0 + 1 + 1 + 1 + 1}$$

$$= 0.59$$

The full similarity matrix is as follows:

		Clone			
		1	2	3	4
Clone	1	1	0.41	0.07	0.40
	2	0.41	1	0.12	0.25
	3	0.07	0.12	1	0.59
	4	0.40	0.25	0.59	1

3.7 Similarity and distance between populations

Until now the discussion has been concerned with the measurement of similarity and distance between individuals, whether organisms or macromolecular sequences. When one moves on to consider the comparison of groups the problems become more difficult. Some of the problems can be summarized by the following headings:

(a) *The choice of a summary statistic for each character to describe a group or population.* This might be a proportion(s) (qualitative characters) or mean value (quantitative characters).

(b) *Measurement of within-group variation.*

(c) *Construction of a measure of similarity or distance* based on (a), and perhaps making allowance for (b). Making allowance for within-group variation might be particularly tricky if this is not constant from one group to another, and there is no reason to believe that it should be.

First similarity and distance measures for qualitative characters will be introduced, and then those for quantitative characters.

3.7.1 Qualitative characters

A distance measure that geneticists have used when describing populations in terms of gene frequencies has the following form.

The *genetic distance*, d_{AB}, between populations A and B is:

$$d_{AB} = (1 - \cos\theta)^{\frac{1}{2}} \qquad (3.16)$$

where

$$\cos\theta = \sum_i (p_{iA} p_{iB})^{\frac{1}{2}} \qquad (3.17)$$

Cos θ is a measure of *genetic similarity* between the two populations, A and B, p_{iA} and p_{iB} being the gene frequencies for the ith allele at a given locus in the two populations. The angular transformation for the proportions has a variance-stabilizing role, and the distance, d_{AB}, in geometrical terms is the chord subtended by the angle θ on a hypersphere of unit radius (see Edwards & Cavalli-Sforza, 1964; and Cavalli-Sforza & Bodmer, 1971, Chapter 11). When several genetic loci are considered the chord lengths for each locus, i.e., the d_{AB} values, are added together. This is equivalent to the construction of a city-block metric.

This approach to measuring distances between populations can be generalized to include data used by numerical taxonomists in the following way. One merely has to replace the word 'locus' in the above definition by 'character', and 'allele' by 'character state'. If there are p qualitative characters there will then be p d_{AB} values

Table 3.6. *Characteristics of two hypothetical populations of red campion*

Character state	Proportion	
	Population A	Population B
Corolla colour		
pink	0.95	0.80
white	0.05	0.20
Coronal scale colour		
as petals, pink	0.85	0.75
not as petals, pink	0.01	0.15
not as petals, white	0.14	0.10
Red calyx pigment		
present	0.80	0.60
absent	0.20	0.40

which can then be added together (in a city-block metric) or squared and then added (to give the square of Euclidean metric). For example, consider the set of proportions from two hypothetical populations of red campion shown in Table 3.6 (cf Table 2.2).

For corolla colour:

$$d_{AB} = [1 - (0.95 \times 0.80)^{\frac{1}{2}} - (0.05 \times 0.20)^{\frac{1}{2}}]^{\frac{1}{2}}$$
$$= 0.17$$

For coronal scale colour:

$$d_{AB} = [1 - (0.85 \times 0.75)^{\frac{1}{2}} - (0.01 \times 0.15)^{\frac{1}{2}} - (0.14 \times 0.10)^{\frac{1}{2}}]^{\frac{1}{2}}$$
$$= 0.21$$

For red calyx pigment:

$$d_{AB} = [1 - (0.80 \times 0.60)^{\frac{1}{2}} - (0.20 \times 0.40)^{\frac{1}{2}}]^{\frac{1}{2}}$$
$$= 0.16$$

The total distance between the two populations is merely the sum of the three distances for the characters taken separately; that is,

$$d_{AB} = 0.17 + 0.21 + 0.16 = 0.54$$

3.7.2   *Quantitative characters*

Perhaps the simplest way to construct a measure of distance between groups is to use their character mean values, and then use a Euclidean or city-block metric. For example, if the two groups had means $\bar{x}_1, \bar{x}_2, \ldots, \bar{x}_p$ and $\bar{y}_1, \bar{y}_2, \ldots, \bar{y}_p$, respectively, for their p characters, we might simply calculate the distance between them as:

$$d_{xy} = \left(\sum_{k=1}^{p} (\bar{x}_k - \bar{y}_k)^2 \right)^{\frac{1}{2}} \tag{3.18}$$

or

$$d_{xy} = \sum_{k=1}^{p} |\bar{x}_k - \bar{y}_k| \tag{3.19}$$

However, it might be more appropriate to consider measures which incorporate, in one way or another, knowledge of intra-OTU variation. One such measure which is being used increasingly in taxonomic studies (see, for example, Fisher, 1969) is Mahalanobis'

D^2 given by

$$D^2 = (\bar{\mathbf{x}} - \bar{\mathbf{y}})'\mathbf{W}^{-1}(\bar{\mathbf{x}} - \bar{\mathbf{y}}) \tag{3.20}$$

where $\bar{\mathbf{x}}$ and $\bar{\mathbf{y}}$ are the *vectors* of means of the two groups; that is, $\bar{\mathbf{x}}' = (\bar{x}_1, \bar{x}_2, \ldots, \bar{x}_p)$, $\bar{\mathbf{y}}' = (\bar{y}_1, \bar{y}_2, \ldots, \bar{y}_p)$, and $\mathbf{W}$ is a $p \times p$ matrix of pooled within-group dispersions for the two groups, i.e. $\mathbf{W} = \mathbf{W}_1 + \mathbf{W}_2$, where $\mathbf{W}_1$ and $\mathbf{W}_2$ are the usual ($p \times p$) matrices of character sums-of-squares and cross-products for the two groups. When correlations between characters are slight, D^2 will be similar to the squared Euclidean distance computed on the standardized data. The use of D^2 implies that the investigator is willing to assume that the character dispersions are at least approximately the same in the two groups. When this is not so D^2 is inappropriate, and in such a case a possible alternative is Jardine and Sibson's Normal Information Radius (see Jardine & Sibson, 1971).

A number of other possibilities exist for between-group distance or similarity measures. For example, the distance between two groups could be defined as the distance between their closest members, one in each group. This is sometimes known as *nearest-neighbour distance* and is the basis of the clustering technique known as *single* linkage, which will be discussed in Chapter 6. A further possibility is the exact antithesis of nearest-neighbour distance, in that the distance between groups is now defined as that between the most remote pair of individuals, one in each group. This is known as *furthest-neighbour distance* and is associated with the *complete* linkage cluster method, also to be described in Chapter 6. Another inter-group measure may be obtained by taking the average of all the inter-individual measures of those pairs of individuals where the members of the pairs are in different groups. This measure is used in *group average clustering* (see Chapter 6). Lance & Williams (1967) point out that the concept of an average for similarity coefficients is not always acceptable and suggest that, where it is unacceptable, a more satisfactory inter-group similarity measure will be obtained from

$$s_{G_1 G_2} = \cos\left(\frac{1}{n_1 n_2} \sum_{\substack{i \in G_1 \\ j \in G_2}} \cos^{-1} s_{ij}\right) \tag{3.21}$$

where $s_{G_1 G_2}$ is the similarity of groups G_1 and G_2, n_1 and n_2 are the number of individuals in these groups, and s_{ij} represents a single inter-individual measure.

3.8   **Summary**

Questions which arise in the measurement of similarity and dissimilarity are numerous and have been discussed only relatively briefly in this chapter. (The issues are more fully discussed in Williams & Dale, 1965; D. G. Morrison, 1967; Sneath & Sokal, 1973; Anderberg, 1973; and Clifford & Stephenson, 1975.) However, the main difficulty for taxonomists is clear: which similarity measure should be used, since different measures may lead to different final classifications? Unfortunately and despite a number of comparative studies (see Cheetham & Hazel, 1969; Boyce, 1969; and Williams, Lambert & Lance, 1966), we are still unable to answer this question in any absolute sense and the choice of coefficient will have to be guided primarily by the type of characters being used and the intuition of the investigator. However, one recommendation made by Sneath & Sokal (1973) which we endorse is to choose the *simplest* coefficient from those applicable to one's data, since this will generally ease the, at times, difficult task of interpretation of final results.

4

Principal components analysis

4.1   Introduction

In the construction of a measure such as Euclidean distance it is assumed that the p characters observed on each OTU may be represented sensibly by p orthogonal axes; however, such an assumption is very rarely true because many of the characters will be *correlated*. For example, Rohlf (1967) describes a study of 40 species of North American mosquitoes in which 74 characters, based on the external morphology of pupae, were used. The characters were mostly counts of the number of branches of various setae in each abdominal segment and lengths of a few setae, and might be expected to be highly correlated. With such data it is frequently useful to attempt to find *uncorrelated* composite measures by the method of *principal components* before attempting any analyses such as clustering, etc., since it is often found that the data may be expressed in terms of far fewer than p of the composite measures without any significant loss of information.

The technique of principal components analysis consists of transforming the set of observed characters, $x_1, x_2, \ldots, x_p$, to a new set, $y_1, y_2, \ldots, y_p$, which has the following properties:

(a) Each y is a linear combination of the xs; that is,

$$y_1 = a_{11}x_1 + a_{12}x_2 + \ldots + a_{1p}x_p$$
$$\vdots$$
$$y_p = a_{p1}x_1 + a_{p2}x_2 + \ldots + a_{pp}x_p$$

(b) The coefficients defining each linear transformation are such that the sum of their squares is unity; that is,

$$\sum_{j=1}^{p} a_{ij}^2 = 1, i = 1, \ldots, p$$

(c) Of all the possible transformations of this type, y_1 has greatest variance.

(d) Of all possible transformations of this type which are uncorrelated with y_1, y_2 has the greatest variance. Similarly y_3 has the greatest variance amongst linear transformations uncorrelated with y_1 and y_2, and so on, until the complete set of p transformed variables has been defined.

Principal components analysis thus leads to a set of p composite characters that are uncorrelated and are arranged in order of decreasing variance. If it is found that the first few principal components (i.e. the y variables) account for most of the variation, it might be possible to use only these in subsequent analyses and thus achieve a considerable simplification.

The next two sections will consider this method of analysis in more detail, beginning in section 4.2 with a geometrical interpretation of the procedure, followed in section 4.3 by a brief mathematical account.

4.2 Principal components analysis – geometrical interpretation

Consider the following hypothetical set of data consisting of pairs of measurements on the flowers from four buttercups:

mean petal length (mm)	8	10	20	30
mean petal width (mm)	4	9	11	18

A plot of these data is shown in Figure 4.1, and clearly the pairs of measurements are highly correlated. If one wished to express the variation in these two characters on a single axis or dimension, what would be the 'best' axis to choose? One intuitively sensible answer to this question would be to choose the axis which *maximizes* the variance of the projections of the four points onto itself, since this will provide the maximum discrimination between the four buttercups. It is easy to show that such an axis is given by the line of best fit, in the least-squares sense, to the points; that is, the line that *minimizes* the sum of squares of the distances between the points and itself. This line is equivalent to the first principal component and is shown in Figure 4.1 for the buttercup data. Projecting the four points onto this line as shown (Figure 4.1(b)) gives the following first *principal component scores* for each buttercup:

9 13 23 35

A line drawn at right angles to the best-fitting line is equivalent to the second principal component axis and the corresponding scores for the four buttercups on this second component are

1   − 1.5   0.5   0

Note that the values for the first principal component are all positive and are clearly related to the size of the buttercup flowers. Those of the second component, however, can have either positive or negative values and may give information about the variation in the shape of the flowers (see section 4.4).

In general, if one has measurements on p variables, the first principal component is the best-fitting straight line in this p-dimensional character space. Similarly, the first two principal components describe the best-fitting plane in this space, and so on. So the best

Figure 4.1. (*a*) Plot of mean petal length against mean petal width for four hypothetical buttercups.
(*b*) Plot of first principal component score derived from measurements of petal length and width.

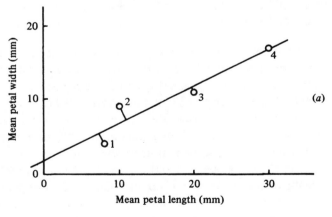

(*a*)

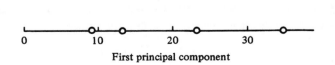

(*b*)

fit in r dimensions (r being less than p) is obtained by projecting the observations into the subspace defined by the first r principal components. How well this r-dimensional configuration of the n taxonomic units describes the configuration in the original p-dimensional space may be measured by the proportion of the variance in the data accounted for by the first r principal components (see next section). To summarize, one may say that a principal components analysis refers the original data to a new set of orthogonal axes. When only the first few principal components axes are used to represent the data, the relative positions of the points in the new character space are an approximation to the relative positions of the OTUs in the original character space, and the Euclidean distance between two points in the principal components space is an approximation to the Euclidean distance between the corresponding points in the original space.

4.3   **A brief mathematical account of principal components analysis**

The first principal component of the observations is that linear combination y_1, of the original variables,

$$y_1 = a_{11}x_1 + a_{12}x_2 + \ldots + a_{1p}x_p \tag{4.1}$$

whose sample variance is greatest for all coefficients, $a_{11}, \ldots, a_{1p}$ (which we may write as the vector $\mathbf{a}_1$). Since the variance of y_1 could be increased without limit simply by increasing the values $\mathbf{a}_1$, a restriction must be placed on these coefficients, and so it is usually required that the sum of squares of the coefficients, i.e. $\mathbf{a}_1'\mathbf{a}_1$, should be set at a value of unity. (The reason for this choice will be indicated later.)

But how useful is this artificial variate constructed from the observed characters? To answer this question one would first need to know the proportion of the total variance attributable to it. If 87% of the variation in an investigation involving six characters could be accounted for by a simple weighted average of the character values, it would appear that almost all of the variation could be expressed along a single continuum rather than in six-dimensional space. This would provide a highly parsimonious summary of the data, which could be useful in later analyses.

The second principal component is that linear combination

$$y_2 = a_{21}x_1 + a_{22}x_2 + \ldots + a_{2p}x_p \tag{4.2}$$

i.e. $y_2 = \mathbf{a}_2'\mathbf{x}$

which has greatest variance subject to the two conditions

$\mathbf{a}_2'\mathbf{a}_2 = 1$ (for the reasons indicated previously)

$\mathbf{a}_2'\mathbf{a}_1 = 0$ (this condition ensures that y_1 and y_2 are uncorrelated)

Similarly, the jth principal component is that linear combination

$$y_j = \mathbf{a}_j'\mathbf{x} \tag{4.3}$$

which has the greatest variance subject to

$\mathbf{a}_j'\mathbf{a}_j = 1$

$\mathbf{a}_j'\mathbf{a}_i = 0$ $i \neq j$

It can be shown that the coefficient vectors $\mathbf{a}_1, \mathbf{a}_2, \ldots, \mathbf{a}_p$ are the *latent vectors* of the *covariance* matrix of the original characters, and that when these are *normalized* or *scaled* so that the sum of their squares is unity, the *latent roots* of this matrix, $\lambda_1, \lambda_2, \ldots, \lambda_p$, are interpretable as the sampling variances of $y_1, \ldots, y_p$, respectively (see D. F. Morrison, 1967). Consequently, we have

$$\lambda_1 + \lambda_2 + \ldots + \lambda_p = s_{11} + s_{22} + \ldots + s_{pp}$$

where $s_{ii}, i = 1, \ldots, p$ are the diagonal elements of $\mathbf{S}$ (the covariance matrix). This may be written more concisely as

$$\sum_{i=1}^{p} \lambda_i = \text{trace}(\mathbf{S}) \tag{4.4}$$

If the first r components explain a large amount of the total variance as indicated by $\sum_{i=1}^{r} \lambda_i / \text{trace}(\mathbf{S})$, then scores on each of these components for each OTU may be used in later analyses in place of the original characters. The *component scores* of the ith OTU are given simply by

$$y_{1i} = \mathbf{a}_1'\mathbf{x}_i$$
$$\vdots$$
$$y_{pi} = \mathbf{a}_p'\mathbf{x}_i \tag{4.5}$$

where $\mathbf{x}_i$ is the vector of character scores for OTU i.

It should be noted that the method of principal components is not independent of the scale(s) of the original measurements. Multiplying one of the variables by a constant (for example, by altering the scale from metres to centimetres) will change the covariance matrix and produce a different set of principal components. It should also be remembered that where the original characters are measured in widely different units, linear combinations of them will have no sensible physical dimensions. Consequently, the analysis is often carried out on standardized measurements and the components extracted from the correlation rather than the covariance matrix. Examples of situations where no standardization is needed, however, include those where all measurements are proportions (for example gene frequencies) or where they are all logarithms of lengths. The effect of the logarithmic transformation in the latter case is to give measurements with the same proportional variability the same variance, so that measurements that are relatively more variable will have a higher variance and will be given more weight in the subsequent analysis.

4.4   Examples

To try to clarify the ideas covered so far a simple example taken from Jolicoeur & Mosimann (1960), which has been previously discussed by D. F. Morrison (1967), will be described. This investigation concerned the measurement of carapace length, width and height in painted turtles. The covariance matrix of these measure-

Table 4.1. *The first three principal components obtained from carapace measurements of painted turtles*

Character	Components (latent vectors of S)		
	1	2	3
Length	0.8126	−0.5454	−0.2054
Width	0.4955	0.8321	−0.2491
Height	0.3068	0.1006	0.9465
Variance (latent roots of S)	680.40	6.50	2.86
Percentage of total variance	98.64	0.94	0.41

Table 4.2. *Coefficients of correlation between winged aphid variables (Jeffers, 1967)*

1	2	3	4	5	6	7	8	9	10	11	12	13	14	15	16	17	18	19
0.934																		
0.927	0.941																	
0.909	0.944	0.933																
0.524	0.487	0.543	0.499															
0.799	0.821	0.856	0.833	0.703														
0.854	0.865	0.886	0.889	0.719	0.923													
0.789	0.834	0.846	0.885	0.253	0.699	0.751												
0.835	0.863	0.862	0.850	0.462	0.752	0.793	0.745											
0.845	0.878	0.863	0.881	0.567	0.836	0.913	0.787	0.805										
-0.458	-0.496	-0.522	-0.488	-0.174	-0.317	-0.383	-0.497	-0.356	-0.371									
0.917	0.942	0.940	0.945	0.516	0.846	0.907	0.861	0.848	0.902	-0.465								
0.939	0.961	0.956	0.952	0.494	0.849	0.914	0.876	0.877	0.901	-0.447	0.981							
0.953	0.954	0.946	0.949	0.452	0.823	0.886	0.878	0.883	0.891	-0.439	0.971	0.991						
0.895	0.899	0.882	0.908	0.551	0.831	0.891	0.794	0.818	0.848	-0.405	0.908	0.920	0.921					
0.691	0.652	0.694	0.623	0.815	0.812	0.855	0.410	0.620	0.712	-0.198	0.725	0.714	0.676	0.720				
0.327	0.305	0.356	0.272	0.746	0.553	0.567	0.067	0.300	0.384	-0.032	0.396	0.360	0.298	0.378	0.781			
-0.676	-0.712	-0.667	-0.736	-0.233	-0.504	-0.502	-0.758	-0.666	-0.629	0.492	-0.657	-0.655	-0.687	-0.633	-0.186	0.169		
0.702	0.729	0.746	0.777	0.285	0.499	0.592	0.793	0.671	0.668	-0.425	0.696	0.724	0.731	0.694	0.287	-0.026	-0.775	

See Table 2.3 for an explanation of the characters 1–19.

ments for 24 female turtles was as follows:

$$S = \begin{bmatrix} 451.39 & & \\ 271.17 & 171.73 & \\ 268.70 & 103.29 & 66.65 \end{bmatrix}$$

The coefficients and variances of the three principal components extracted from this matrix are shown in Table 4.1. Note that the sum of squares of the coefficients in each component is unity and that the sum of the latent roots is equal to the sum of the diagonal elements in S.

The first principal component accounts for nearly all of the variance in the three characters. It is a form of weighted mean of the carapace measurements

$$y_1 = 0.81\text{(length)} + 0.50\text{(width)} + 0.31\text{(height)}$$

The size of the turtle shells could be characterized by this single variable with little loss of information since it alone accounts for some 98% of the variation of the three measurements length, width and height. However, it is of some interest to examine the remaining two components

$$y_2 = -0.54\text{(length)} + 0.83\text{(width)} + 0.10\text{(height)}$$

and $$y_3 = -0.20\text{(length)} - 0.25\text{(width)} + 0.94\text{(height)}$$

both of which appear to be measures of carapace 'shape', being comparisons of length versus width and height, and height versus length and width, respectively. The first principal component often has the characteristic of a measurement of size. Jolicoeur & Mosimann (1960) emphasize that for this interpretation to be justified all coefficients must have the same sign, whereas those of the other components must generally have mixed signs. Rao (1964) gives a mathematical argument for this interpretation, and the interested reader is referred to Blackith & Reyment (1971) for a fuller discussion of this point.

As a second example, a set of data described by Jeffers (1967) of 40 individual winged aphids will be considered. Measurements on 19 morphological variables were obtained (see Table 2.3), and the correlations between these measurements are shown in Table 4.2. The high degree of correlation between almost all of the variables

Table 4.3. *The latent roots for the first four principal components obtained from the aphid data* (*Table* 4.2)

Component	Latent root	Percentage of variability	
		Component	Cumulative
1	13.86	73.0	73.0
2	2.37	12.5	85.5
3	0.75	3.9	89.4
4	0.50	2.6	92.0

Table 4.4. *Latent vectors for first four components of winged aphid variables* (*Jeffers*, 1967)

Variable	Latent vectors for component			
	1	2	3	4
Length	0.96	−0.06	0.03	−0.12
Width	0.98	−0.12	0.01	−0.16
Fore-wing	0.99	−0.06	−0.06	−0.11
Hind-wing	0.98	−0.16	0.03	−0.00
Spiracles	0.61	0.74	−0.20	1.00
Antennal segment I	0.91	0.33	0.04	0.02
Antennal segment II	0.96	0.30	0.00	−0.04
Antennal segment III	0.88	−0.43	0.06	−0.18
Antennal segment IV	0.90	−0.08	0.18	−0.01
Antennal segment V	0.94	0.05	0.11	0.03
Antennal spines	−0.49	0.37	1.00	0.27
Tarsus III	0.99	−0.02	0.03	−0.29
Tibia III	1.00	−0.05	0.09	−0.31
Femur III	0.99	−0.12	0.12	−0.31
Rostrum	0.96	0.02	0.08	−0.06
Ovipositor	0.76	0.73	−0.03	−0.09
Ovipositor spines	0.41	1.00	−0.16	−0.06
Fold	−0.71	0.64	0.04	−0.80
Hooks	0.76	−0.52	0.06	0.72

indicates that most of the variation might be explained by only a few principal components, and this view is confirmed by the results in Table 4.3. The coefficients for the first four principal components are given in Table 4.4. Note that the latent vectors have been rescaled by Jeffers after the analysis, so that the maximum coefficient (weighting) in each column is $+1$. This scaling is arbitrary and is only useful if it helps the reader to assess the relative importance of each of the original measurements in the composition of the calculated orthogonal components. In this example the first component is again clearly a measure of the overall size of the aphids. Jeffers (1967) interpreted the second component as a measure of the number of ovipositor spines, the third as a measure of the number of antennal spines and the fourth as a measure of the number of spiracles.

4.5 Principal components plots

The principal components scores that can be obtained for each OTU might be useful in replacing the original character values before the computation of an inter-OTU proximity matrix and in subsequent analyses such as clustering, etc. (see Chapter 6). However, the scores may also be used in a more immediate fashion by making principal component scatter plots, thus enabling a direct visual examination of the relationships between the OTUs to be made. Of course, with data involving only two character values for each OTU, such as the buttercup data in section 4.2, it is possible to obtain a visual representation simply by a plot of the raw data; however, if the data are more complex (as they generally are!), it is impossible to visualize a p-dimensional scatter plot of the original measurements, and so plotting the points along the first principal component, or plotting the first component score for each OTU against the second, may be very informative, particularly when these first components account for a large proportion of the variance. An example of such a plot for the aphid data appears in Figure 4.2. Here four fairly distinct groups or clusters of aphids can be seen. When it is thought that the first two components are inadequate for representing the structure in the data, a further component score can be accommodated by making a three-dimensional 'ball and wire' model or by constructing stereographic plots (see Rohlf, 1968, for examples). (A number of other simple methods which enable more than two scores to be plotted on a two-dimensional diagram are described in Everitt (1978).)

In many taxonomic studies it is variation in shape rather than

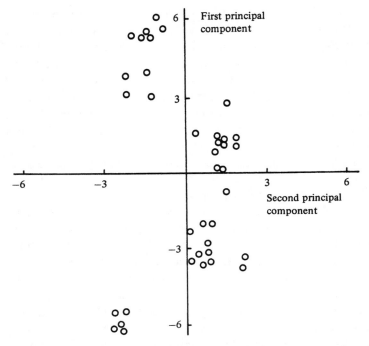

Figure 4.2. First two principal component scores for winged aphids (Taken with permission from Jeffers, 1967.)

variation in size which is of interest; consequently, if it is thought that the first principal component is a measure of size, as in the cases of the turtles and aphids discussed previously, then more information about the *relevant* structure amongst the OTUs might be obtained by plotting the second principal component against the third, rather than a plot of components one and two. For example, Temple (1968) found that such a plot (see Figure 4.3) clearly revealed the presence of two forms in the Silurian brachiopod, *Toxorthis*. (Of course, a three-dimensional model constructed from components one, two and three would equally have revealed this heterogeneity.)

Although principal components analysis is essentially intended for use with measurements made on a continuous scale, Gower (1966) shows that it may also legitimately be employed on binary data, although the interpretation may be difficult. In particular, the principal components plots arising from such data often suffer from

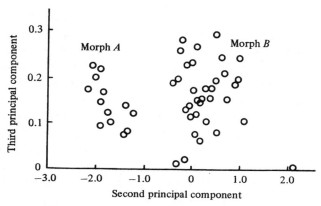

Figure 4.3. Second two principal component scores for the two morphs of *Toxorthis proteus* (Redrawn from Temple, 1968.)

what is generally termed the *horseshoe effect* (see Kendall, 1971, 1975). This effect has been noted in ecological data by Swann (1970), Austin (1972) and others. It refers to the tendency of the points in the plot to lie on a horseshoe-shaped curve, rather than indicate any real structure in the data. The effect arises because presence–absence data occupy some of the apices of a hypercube. A possible method for overcoming this problem is described by Williamson (1978).

It should be remembered that in principal components plots the Euclidean distance between two points representing two OTUs acts as an approximation to the Euclidean distance between these OTUs in the original p-dimensional character space. This implies that such plots are only likely to be useful when the appropriate metric for the OTUs is considered to be Euclidean. In other cases one of the methods to be considered in the next chapter might be more suitable for obtaining informative visual representations of the data.

4.6 Factor analysis

It would be inappropriate to end this chapter without some mention of the term *factor analysis*, since this does occasionally arise in numerical taxonomic exercises (see, for example, Fisher, 1968). Essentially, factor analysis, like principal components analysis, is a technique for data reduction. However, whereas the latter is simply the transformation of the coordinate axes of a multivariate system to new orientations through the natural shape of the scatter swarm of the observations, factor analysis proposes a fundamental model for the covariance structure of the observations. In essence, under the factor model each observed variable is postulated to be a linear function of a small number of unobservable *common-factor* variates and a single latent *specific* variate. The common factors generate the covariances among the observed variables, while the specific terms contribute only to the variances of their particular variables. Such models are widely used in the behavioural sciences where they have largely overcome their 'black sheep' image, obtained during the 1930s. However, in biological work in general and numerical taxonomy in particular there appears to be little to recommend factor analysis over the alternatives such as principal components analysis and the methods to be discussed in Chapter 5. Consequently, no details of the method will be given here and the interested reader is referred to the book by Bennett & Bowers (1976).

4.7   Summary

Principal components analysis has been used widely in numerical taxonomic studies. It can be very useful for displaying the relationships between the OTUs in a low-dimensional space, thus enabling direct visual examination of the relative positions of the OTUs. It is, however, only suitable for this purpose when a Euclidean metric is considered suitable for the observations. In other cases there are more appropriate techniques which can be applied to obtain informative low-dimensional plots, and these will be the subject of the next chapter.

5

Multidimensional scaling

5.1   Introduction

Having calculated a matrix of similarities or dissimilarities between the OTUs, the numerical taxonomist has two main possibilities for its analysis. The first is to apply some form of *cluster analysis*, this being a generic term for methods which seek to determine homogeneous subsets of OTUs and therefore produce a classification of these directly. The second is to use some type of *multidimensional scaling* or *ordination* technique which represents the OTUs as points in some, one hopes, low-dimensional space, in which Euclidean distances between points will reflect the relationship between the corresponding OTUs as indicated in the observed proximity matrix. (Note that multidimensional scaling can just as easily be used for populations, given an appropriate distance measure, as for individual organisms.) When a satisfactory representation is obtained in a space of two or three dimensions, diagrams and models may be constructed which allow a visual assessment of the relationships between the OTUs and the informal examination of their structure, although no classification as such is produced. It is these techniques that will be the subject of this chapter; clustering methods will be discussed in Chapter 6.

5.2   Classical multidimensional scaling

One example of an ordination technique has already been described in the previous chapter. The plotting of a few components can give a low-dimensional representation of the data in which the Euclidean distance between two points acts as an approximation to the *Euclidean* distance between the corresponding OTUs in the original character space. Multidimensional scaling in general is concerned with the problem of constructing a configuration of points in Euclidean space which reflects, in some sense, the relationships between a set of OTUs as implied by their observed proximities *whether these are Euclidean or not.* The classical solution to this problem which operates directly on the observed proximity matrix is commonly known as *principal co-ordinates analysis.* (In the case where the observed proximities are Euclidean distances the results are equivalent to those obtained from a principal components analysis.)

Consider, for example, the results given in Table 3.4(*b*). Here the distances between the three cytochrome *c* peptide sequences might be best represented by the city-block metric based upon minimum mutation values. If one wished to illustrate these distances graphically, one could easily use a pair of compasses to produce a diagram like that in Figure 5.1. Note that the city-block distances in eight-dimensional space (only eight amino-acid positions are being

Figure 5.1. Euclidean representation of the city-block distances given in Table 3.4 (*b*)

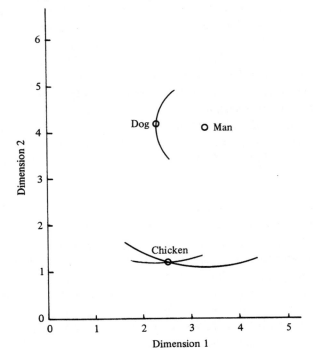

considered) are perfectly represented by the Euclidean distances of the two-dimensional plot. This is because only three OTUs are being considered, and they are bound to be coplanar. If, however, there had been results for several OTUs, there would inevitably have been some distortions in the two-dimensional representation. The aim of principal coordinates analysis is to produce a Euclidean representation of the observed distances that minimizes this distortion; that is, to produce a pattern of points that best represents the pattern in the original multidimensional character space.

5.2.1 *Principal coordinates analysis – technical details*

Let A be a symmetric $(n \times n)$ matrix with latent roots $\lambda_1, \lambda_2, \ldots, \lambda_n$ and associated $(n \times 1)$ latent vectors $c_1, c_2, \ldots, c_n$ as shown in Table 5.1. Suppose we now regard the elements of the ith row of this table as the coordinates of a point in n-dimensional space. The squared Euclidean distance, d_{ij}^2, between two points i and j in this space is, consequently,

$$d_{ij}^2 = \sum_{k=1}^{n} (c_{ik} - c_{jk})^2 \tag{5.1}$$

$$= \sum_{k=1}^{n} c_{ik}^2 + \sum_{k=1}^{n} c_{jk}^2 - 2 \sum_{k=1}^{n} c_{ik}c_{jk}$$

If the latent vectors are normalised so that the sums of squares of their elements are equal to their corresponding latent roots; that is,

Table 5.1. *Latent roots and vectors of the symmetric matrix* A

	Latent vectors			
	c_1	c_2	. . .	c_n
	c_{11}	c_{12}	. . .	c_{1n}
	c_{21}	c_{22}	. . .	c_{2n}
	.	.		.
	.	.		.
	.	.		.
	c_{n1}	c_{n2}	. . .	c_{nn}
Latent roots:	λ_1	λ_2	. . .	λ_n

so that

$$\sum_{k=1}^{n} c_{ki}^2 = \lambda_i \qquad (5.2)$$

then it is well known that

$$\mathbf{A} = \mathbf{c}_1\mathbf{c}_1' + \mathbf{c}_2\mathbf{c}_2' + \ldots + \mathbf{c}_n\mathbf{c}_n' \qquad (5.3)$$

and therefore that

$$a_{ii} = \sum_{k=1}^{n} c_{ik}^2 \quad \text{and} \quad a_{ij} = \sum_{k=1}^{n} c_{ik}c_{jk}$$

Consequently

$$d_{ij}^2 = a_{ii} + a_{jj} - 2a_{ij} \qquad (5.4)$$

Therefore, if $\mathbf{A}$ is a similarity matrix with self-similarities taking the value unity, then taking the elements of the latent vectors of $\mathbf{A}$, suitably scaled, as defining coordinate values in n-dimensional space, leads to a representation of the OTUs in which the Euclidean distance, d_{ij}, between points i and j representing OTUs i and j, having similarity s_{ij}, takes the value

$$d_{ij} = [2(1 - s_{ij})]^{\frac{1}{2}} \qquad (5.5)$$

Consequently, OTUs with high similarity values will be represented by points close together in the n-dimensional space and vice versa, which is obviously what we require. However, the method has, as yet, only led to an n-dimensional representation for which there is a sensible relationship between inter-OTU similarities and inter-point distances. What we would like is a corresponding representation in far fewer dimensions. Gower (1966) shows that this may be obtained simply by referring these n-dimensional coordinates to their principal axes (cf. principal components analysis) and using the projected coordinates to display the relationships between the OTUs as implied by their similarities.

Returning for the moment to (5.4), let us now consider what would happen if $\mathbf{A}$ was a dissimilarity matrix with elements δ_{ij} and diagonal entries of zero. Formula (5.4) now becomes

$$d_{ij}^2 = -2\delta_{ij} \qquad (5.6)$$

Consequently the principal coordinates procedure applied directly to a dissimilarity matrix would not produce sensible results! However, if A contained values $a_{ii} = 0$ and $a_{ij} = -\frac{1}{2}\delta_{ij}^2$, then (5.4) becomes

$$d_{ij}^2 = \delta_{ij}^2 \qquad (5.7)$$

and principal coordinates analysis leads to a set of coordinates with Euclidean distances equal to the observed dissimilarities. Again the n-dimensional coordinates may be referred to their principal axes to obtain a representation in fewer dimensions. (When the observed dissimilarities are Euclidean distances principal coordinates analysis and principal components analysis are equivalent.)

Gower (1966) shows that a given similarity or dissimilarity measure may be represented in a Euclidean space in this way only if the matrix α with the following elements has no negative latent roots:

$$\alpha_{ij} = a_{ij} - \bar{a}_{i.} - \bar{a}_{.j} + \bar{a}_{..} \qquad (5.8)$$

where $\quad \bar{a}_{i.} = \dfrac{1}{n} \sum_{j=1}^{n} a_{ij},$

$$\bar{a}_{.j} = \frac{1}{n} \sum_{i=1}^{n} a_{ij}$$

and $\quad \bar{a}_{..} = \dfrac{1}{n^2} \sum_{i=1}^{n} \sum_{j=1}^{n} a_{ij}$

However, a useful approximate representation may still be obtained in cases where there are a *small* number of *small* negative roots. A measure of the adequacy of fit given by principal coordinates analysis is based on the latent roots of the matrix α. For the first r principal coordinates this is given by

$$T = \sum_{i=1}^{r} \gamma_i \Big/ \operatorname{trace}(\alpha) \qquad (5.9)$$

where the γ_i are the latent roots of the matrix α arranged in descending order of magnitude. (When α has a number of negative latent roots Mardia, Kent & Bibby (1979) suggest taking as a goodness of fit measure $\sum_{i=1}^{r} \gamma_i / \sum_{i=1}^{n} |\gamma_i|$.)

This has been only a brief account of the method of principal coordinates; fuller accounts are given by Gower (1966), Blakith &

Table 5.2. *Distance matrix for 28 wheat varieties (Baum, 1977)*

OTU	1	2	3	4	5	6	7	8	9	10	11	12	13	14	15	16	17	18	19	20	21	22	23	24	25	26	27
2	911																										
3	806	936																									
4	950	711	962																								
5	969	835	923	788																							
6	959	877	921	810	665																						
7	959	877	948	875	833	831																					
8	975	928	923	946	876	863	875																				
9	900	912	830	926	886	836	924	814																			
10	964	902	902	864	862	826	927	825	710																		
11	891	897	858	934	840	856	910	860	759	865																	
12	951	765	867	792	728	800	784	857	838	826	902																
13	926	756	903	711	766	790	875	881	871	794	889	779															
14	900	815	807	845	754	800	822	901	849	869	898	688	723														
15	884	810	834	907	844	857	904	919	809	892	767	849	889	832													
16	850	947	764	964	943	921	940	882	811	906	862	899	916	886	862												
17	934	891	865	931	921	886	849	817	841	865	843	883	935	934	830	876											
18	980	917	956	936	931	938	914	863	905	908	942	857	949	914	935	934	894										
19	967	864	742	786	808	829	890	848	863	882	765	781	822	885	945	936	914	935									
20	959	791	946	640	818	805	954	929	882	945	803	815	701	816	945	953	943	955	792								
21	915	782	864	815	899	811	882	887	828	778	866	752	832	815	803	880	744	937	858	825							
22	913	883	903	873	875	810	888	699	698	790	846	816	881	884	814	888	809	799	896	831	790						
23	933	832	841	809	764	781	823	916	866	879	883	717	678	543	872	869	939	938	805	719	833	902					
24	941	881	869	875	846	884	901	901	865	888	752	876	908	934	818	896	786	831	889	913	835	855	908				
25	939	877	870	877	894	877	875	905	767	743	869	758	800	805	853	840	879	919	848	912	814	762	824	913			
26	955	872	924	872	924	870	917	868	710	713	834	851	893	842	880	881	853	932	880	841	892	779	866	915	783		
27	928	928	872	883	806	738	896	872	660	650	859	862	802	851	855	805	893	841	893	841	849	754	842	851	887	746	
28	736	921	703	909	908	919	933	944	842	896	781	904	879	852	824	829	864	941	922	859	934	926	854	819	897	918	881

Reyment (1971), and Mardia *et al.* (1979); here we move on to illustrate the use of the method in taxonomy with an example.

5.2.2   Principal coordinates analysis—an example

Since the results of a principal coordinates analysis often look superficially similar to those of a principal components analysis, it would add little to this discussion to describe an example whose interpretation was quite straightforward. Instead, an example illustrating potential pitfalls will be introduced. The most common of these occurs with data that cannot be summarized adequately in terms of the positions of the OTUs in a two- or three-dimensional space; that is, the first two or three latent roots obtained from a principal coordinates analysis or a principal components analysis do not explain most of the variation in the original character space. This might be the case where one had selected, by accident or by design, a set of characters whose states were all more or less uncorrelated. In this situation, one would be liable to make unjustified inferences from inspection of a two- or three-dimensional principal coordinates (or principal components) plot.

The following example, from Baum (1977), illustrates some of the problems caused by this situation. The dissimilarity matrix shown

Table 5.3. *Latent roots computed from the dissimilarity matrix of Table 5.2 (Baum, 1977)*

Dimension	Latent root	Dimension	Latent root
1	1.217 16	15	0.268 666
2	0.942 037	16	0.254 098
3	0.722 454	17	0.238 131
4	0.598 507	18	0.234 938
5	0.558 098	19	0.215 687
6	0.538 583	20	0.208 695
7	0.460 580	21	0.204 331
8	0.422 575	22	0.179 981
9	0.401 011	23	0.165 221
10	0.386 413	24	0.159 988
11	0.360 360	25	0.150 528
12	0.318 351	26	0.133 011
13	0.301 592	27	0.118 174
14	0.277 315		

in Table 5.2 is from a study of 28 wheat varieties. Unfortunately, the author did not state what, or how many, characters had been used in this study, or how the dissimilarity matrix had been constructed. For the present argument, however, these points are relatively unimportant. This matrix, when subjected to a principal coordinates analysis, yields latent roots that only gradually decrease in size (Table 5.3). The first three latent roots account for roughly 28% of the total variation, and so a plot of the points in the space of the first two or three principal coordinate axes might be expected to distort greatly the relationships between the OTUs, as implied by the observed dissimilarities. For example, consider Figure 5.2(*a*). Here are plotted the coordinates of the first two principal axes (explaining 21.47% of the total variation). The OTUs appear to be clustered into perhaps four groups, but the possible inadequacies of a two-dimensional representation for these data should lead us to question whether these clusters are 'real' or simply artefacts. Baum's further analyses suggest the latter.

To overcome this problem Baum suggested transforming the original dissimilarities in such a way as to maximize the proportion of the total variation explained by the first two or three latent roots

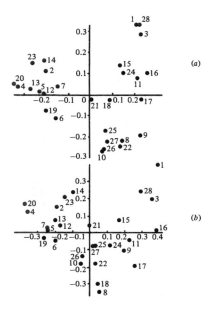

in any subsequent principal coordinates analysis. He found that the transformation

$$\delta'_{ij} = \delta^t_{ij} \tag{5.10}$$

was promising, where the δ_{ij} are the original dissimilarities, the δ'_{ij} the transformed values, and t was chosen empirically to take the value 6. (A summary of the results that led to this choice for t appear in Table 5.4.) A two-dimensional principal coordinates plot obtained after taking this transformation is shown in Figure 5.2(*b*). Here the first two latent roots explain 72.84% of the total variation in the transformed dissimilarity matrix. In this figure there appears to be no obvious clustering of the OTUs, suggesting that the distinct groups indicated by Figure 5.2(*a*) might indeed be artefacts produced by an inadequate analysis.

It might be felt that it is difficult to justify a transformation of the above type even though it might produce informative results. An alternative approach is to consider only the *rank* of the original dissimilarities rather than the dissimilarities or transformed dissimilarities. (The transformation described above is rank preserving.) One can also use different criteria to measure the distortion produced

Figure 5.2. (*a*) Principal coordinates of raw wheat data.
(*b*) Principal coordinates of transformed wheat data (Taken with permission from Baum, 1977.)
(*c*) Non-metric multidimensional scaling of raw wheat data.

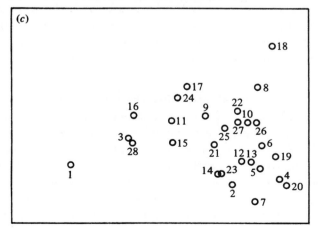

Table 5.4. *Effect of transformation of distance measures on the results of principal coordinates analysis of wheat data (Baum, 1977)*

Transformation	First three latent roots	Cumulative percentage of total variation
X^2	1.48226	19.54
	1.06669	33.60
	0.77173	43.77
X^3	1.54122	26.28
	1.06210	44.39
	0.75215	57.21
X^4	1.49840	32.51
	1.00542	54.32
	0.69629	69.43
X^5	1.40880	38.34
	0.93338	63.74
	0.62490	80.74
X^6	1.30090	43.85
	0.86013	72.84
	0.55078	91.41

by any reduction in dimensionality. *Non-metric multidimensional scaling* (to be discussed in detail in the next section) is an alternative to principal coordinates analysis which utilizes only the rank order of the dissimilarities. The result of an analysis of this type on the current example is given in Figure 5.2(*c*). Note that the result, although not identical, is very close to that obtained from a principal coordinates analysis on the transformed data. This supports the view that non-metric multidimensional scaling might be the most informative and convenient of the two methods of ordination.

5.3   Other methods of multidimensional scaling

The central motivating concept of multidimensional scaling is that the distances between the points representing the OTUs should correspond in some sensible way to the observed proximities. (In principal coordinates analysis, for example, we have that distance = dissimilarity.) With this in mind, various authors, for example, Shepard (1962), Kruskal (1964*a*) and Sammon (1969), have approached the problem by defining an objective function which measures in some way the discrepancy between the observed proximities and the fitted distances, and then attempted to recover the configuration in a particular number of dimensions which minimizes this function using one of the many optimization algorithms now available. For example, let us suppose that the proximity matrix under investigation contains a measure of dissimilarity for each pair of OTUs. We require a set of r-dimensional coordinates (where we hope that r is of the order 2 or 3), with associated Euclidean distances, d_{ij}, which match in some way the observed dissimilarities. To assess the agreement between distances and dissimilarities we need to define some function that takes the value zero if the pattern of the distances fits that for the dissimilarities perfectly in some sense, and increases in value as the fit becomes less good. An intuitively obvious candidate is the sum of squares criterion, SS, given by

$$\text{SS} = \sum_{i=1}^{n} \sum_{j=i+1}^{n} (\delta_{ij} - d_{ij})^2 \tag{5.11}$$

Since the distances d_{ij} are a function of the n r-dimensional coordinate values, so also is SS, and by changing the coordinate values we change the value of SS. Since goodness-of-fit increases with decreasing values of SS, we would now seek to determine that set of r-dimensional coordinates that *minimizes* SS. Various optimization algorithms, such as steepest descent (see Kruskal, 1964*b*), might be considered, but the details of these need not concern us here.

For various reasons, the simple sum of squares criterion given in (5.11) is rarely used directly as a measure of goodness-of-fit of a set of coordinates, but is first 'weighted' or otherwise altered before being minimized. For example, Sammon (1969) uses the function

$$S_1 = \frac{1}{\sum_{i<j} \delta_{ij}} \sum_{i<j} \frac{(\delta_{ij} - d_{ij})^2}{\delta_{ij}} \tag{5.12}$$

Since OTUs with high dissimilarities contribute little to S_1, they may be represented by points having distances which are a very poor match for the observed dissimilarity. On the other hand, OTUs with low dissimilarities should be well fitted in the final configuration. Consequently, S_1 tends to preserve *local* structure. (This should be contrasted with the results of a classical technique such as principal

components analysis, which are characterized by a good representation of large distances, but are poor at representing distances
between close neighbours; see Rohlf (1968).)

Of particular interest in numerical taxonomy, where an observed
proximity might only be regarded as an *indication* of the relative
similarity or dissimilarity between two OTUs, with its actual
numerical value being of little interest, is *non-metric multidimensional
scaling*. This is used in an increasing number of taxonomic investigations, as we shall see in section 5.3.2 after a brief technical
description of the method.

5.3.1   *Non-metric multidimensional scaling – technical details*

To introduce this method let us consider a situation where
we have four OTUs and six observed values of dissimilarity for the
six possible pairs. Now suppose that the following rank ordering
holds for these six dissimilarity values:

$$\delta_{23} < \delta_{12} < \delta_{34} < \delta_{13} < \delta_{24} < \delta_{14} \tag{5.13}$$

In other words, the second and third OTUs are judged to be least
dissimilar (or most similar), the first and second OTUs next least
dissimilar, and so on, with OTUs 1 and 4 ranked as most dissimilar
(or least similar). Now let the OTUs be represented as points in a
Euclidean space of a specified dimensionality with an associated set
of distances, $d_{12}, d_{13}, d_{14}, d_{23}, d_{24}, d_{34}$. In non-metric multidimensional scaling these distances are considered to match the
observed dissimilarities *perfectly* if they satisfy the following
relationship:

$$d_{23} \leq d_{12} \leq d_{34} \leq d_{13} \leq d_{24} \leq d_{14} \tag{5.14}$$

That is, the order relationship among the inter-point distances in
the Euclidean representation of the OTUs is in exact concordance
with the observed relationship among the observed dissimilarities.
In other words, the distances are *monotonic* with the dissimilarities.

Such a perfect match may, of course, not hold in a particular
Euclidean representation and so a measure to evaluate the fit of any
given configuration to the monotonicity requirement is needed. For
this purpose Kruskal (1964*a*) defines a function called *stress* given by

$$\text{stress} = \left[\sum_i \sum_j (d_{ij} - \hat{d}_{ij})^2 \Big/ \sum_i \sum_j d_{ij}^2 \right]^{\frac{1}{2}} \tag{5.15}$$

Again this measure is essentially a sum of squares; it is a function of the coordinates of the points used to represent the OTUs (since the distances, d_{ij}, depend on these coordinate values). In this formula the $\hat{d}_{ij}$ are a set of numbers known to be monotonic with the dissimilarities. They are *not* distances; there is no configuration whose inter-point distances are $\hat{d}_{ij}$. They are merely a set of numbers monotonic with the δ_{ij}, used to reference the departure from monotonicity of the fitted distances, d_{ij}. Stress therefore measures the extent to which those distances are monotonic with the observed dissimilarities, low values of stress indicating a high degree of concordance of the rank orderings of dissimilarities and distances.

Having defined a suitable goodness-of-fit measure for the required criterion of fit, it now only remains to find that set of coordinates in a specified number of dimensions minimizing the measure. Again this may be done with a variety of optimization algorithms (for details, see Kruskal, 1964*b*; and Kruskal & Wish, 1978). The most important thing to note with this approach to multidimensional scaling is that the actual numerical values of the dissimilarities are *not* used; only their rank order is of importance and so the method would give the same results under any *monotonic* transformation of the dissimilarity values.

5.3.2 *Non-metric multidimensional scaling – examples*

In this section two examples are described which illustrate the type of information that may be gained from an inspection of two-dimensional scatter plots. The plots are the results of non-metric multidimensional scaling, using the method described by Sibson (1972), but the discussion is also applicable to the interpretation of the results of other types of ordination.

The first example (Figure 5.3) shows variation between 97 European populations of white campion (*Silene alba* (Miller) E. H. L. Krause) and red campion (*Silene dioica* L. Clairv.). The second (Figure 5.4) shows variation between 29 Scottish populations of red campion. They are taken from a study of the taxonomy of the genus *Silene* by Prentice (1979, 1980), the original dissimilarity matrices being constructed using Jardine and Sibson's information radius (or *K*-dissimilarity).

Consider Figure 5.3. The 57 white campion populations are clearly distinct from the 40 of the red campion, indicating, at least, that the primary classification of these populations into two species is reasonably sensible! One can also clearly see from this scatter plot that the white campion is much more variable than the red campion. Now consider Figure 5.4. Here the shape of the symbols indicates the geographical location of the OTUs, and the accompanying

Figure 5.3. Multidimensional scaling of 97 *S. alba* and *S. dioica* populations (seed, flower and capsule characters). Most of the 40 *S. dioica* populations were positioned within a dense cluster, indicated by the black region. (Taken with permission from Prentice, 1979.)

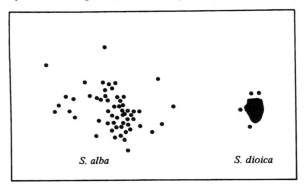

Figure 5.4. Multidimensional scaling of 29 Scottish populations of *S. dioica*. Symbols indicate geographic areas: closed squares, S. Scotland; closed triangles, W. Ross; closed circles, Skye; open squares, Shetland; open triangles, N. Scotland; open circles, C./E. Scotland. Numbers indicate habitat types: 1, woodland; 2, hedgerow or roadside bank; 3, coastal cliff; 4, saltmarsh. (Taken with permission from Prentice, 1980.)

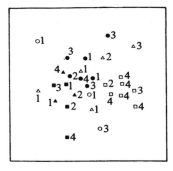

numbers their habitat. There appears to be no obvious clustering within this subset of red campion populations, but there is evidence of geographical differentiation. These two scatter plots show clearly how ordination can be used as a powerful method for summarizing a large amount of data. They can be used to suggest ideas, and to indicate whether the results of a cluster analysis, to be described in the next chapter, are likely to be sensible.

5.4   Minimum spanning trees

The usefulness of the techniques discussed in the previous sections must be judged by how well the inherent structure in the data is preserved by the ordination procedure. Essentially this will depend on how well the original proximities are preserved by the mapping process, and in part this may be judged by the goodness-of-fit criteria already mentioned in connection with particular techniques. However, other methods are available for comparing two sets of distances which are often very useful. For example, Sokal & Rohlf (1962) and Kruskal & Carroll(1969) suggest correlating the $n(n-1)/2$ distance pairs, (d_{ij}, d_{ij}^*), where the d_{ij}s are the original distances between the OTUs and the d_{ij}^*s are the distances between points representing the OTUs and resulting from some particular ordination process. This correlation coefficient is generally known as the *cophenetic correlation*, and will be discussed in detail in the next chapter.

A method which is particularly suitable for assessing how well the original proximities are preserved by a two-dimensional mapping uses the *minimum spanning tree* of a proximity matrix. This may be defined as follows:

Suppose n points are given (possibly in many dimensions); the *tree* spanning these points, i.e. *a spanning tree*, is any set of straight-line segments joining pairs of points such that

- (*a*) No closed loops occur
- (*b*) Each point is visited by at least one line
- (*c*) The tree is connected; that is, it has paths between any pair of points.

If a weight is assigned to each segment in the tree then its *length* is defined to be the sum of these weights. The *minimum spanning tree* of the *n* points is then defined as the spanning tree of minimum length. These concepts are illustrated in Figures 5.5(*a*) and 5.5(*b*).

The links of the minimum spanning tree of the original proximity matrix may be plotted onto the two-dimensional representation obtained from a particular ordination technique, and the resulting diagram may, as the example below illustrates, help to highlight distortions produced by the mapping process. (Algorithms to find the minimum spanning tree are described by Prim (1957), Gower & Ross (1969) and Farris (1970); see Chapter 8.)

To illustrate the use of the minimum spanning tree for indicating 'inadequacies' in two-dimensional representations of higher dimensional data we shall use the example described by Gower & Ross (1969). This involved ten skull measurements of white-toothed shrews from the Scilly and Channel Islands. A distance matrix between ten island races of such shrews was derived, based upon the ten skull measurements. This was then represented in two dimensions by means of a canonical variate analysis (see Chapter 7), with the result

Figure 5.5. Two spanning trees for a set of six points: (*a*) minimum spanning tree of length 20.6; (*b*) spanning tree of length 27.8.

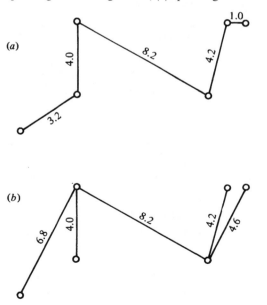

shown in Figure 5.6(*a*). This two-dimensional representation accounts for 89% of the variance, but that there is still some distortion is readily apparent from examining Figure 5.6(*b*), which shows the minimum spanning tree of the original distance matrix superimposed upon the two-dimensional mapping. In Figure 5.6(*a*), the Jersey and Sark races appear to be well separated from those of the other

Figure 5.6. (*a*) Canonical variate plot of shrew data.
(*b*) Canonical variate plot with minimum spanning tree superimposed. (Taken with permission from Gower and Ross, 1969.)

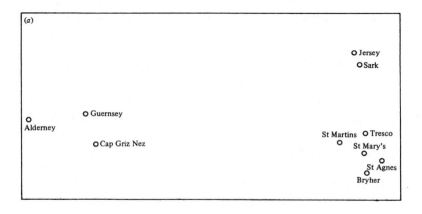

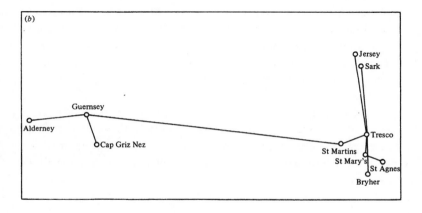

Channel Islands and also from the Scilly Island races. However, Figure 5.6(*b*) indicates that the Jersey and Sark races are *closer* to the Tresco race than to each other (otherwise the Jersey–Sark join would be a link in a shorter tree), so that the separation from those of the Scilly Islands is illusory. The tree also shows that the Bryher race is closer to that of Tresco than to those of St Mary's and St Agnes, so that the Scilly Island cluster is more compact than it appears in Figure 5.6(*a*).

5.5   Summary

The techniques described in this chapter make it possible to represent the relationships between a set of OTUs, as indicated by their observed proximities, by the Euclidean distances between a set of points in some specified number of dimensions. When an adequate fit is obtained in two or three dimensions diagrams and models may be constructed which allow a visual examination of these relationships and these may be extremely valuable in taxonomic investigations. In general, the choice of a particular method from amongst those discussed in this chapter will be governed by many factors, for example, the type of characters observed, the number of OTUs involved, the complexity of the structure expected, and so on. No convincing case can be made for the use of one specific method in all situations; indeed it may not be unreasonable to use more than one of the techniques in many cases. For many sets of data the different methods may all give very similar results. However, since in general fewer dimensions will be needed to reproduce ordinal information than to reproduce metric information, non-metric multidimensional scaling techniques are perhaps more likely to give useful and informative low-dimensional representations of the data than other methods, particularly when any structure is of a complex non-linear variety, for example, points scattered around a curve in *p* dimensions.

The techniques discussed in this chapter do not produce a classification of the OTUs *per se*, although they may be very suggestive of the appropriate groupings. In the next chapter methods which are designed to construct classifications directly will be described.

6

Cluster analysis

6.1 Introduction

The ordination or scaling techniques described in the last two chapters may be very useful for indicating the taxonomic structure in a collection of organisms. However, they do not lead to an *explicit* separation of the organisms into groups and so neither do they produce a classification *per se*. For this the numerical taxonomist turns to one of the many available methods of *cluster analysis*. Such techniques have developed rapidly over the last two decades and there is now a vast body of literature associated with them. In a single chapter, in an introductory text such as this, it will be possible to deal with only a small number of the methods that have been suggested, concentrating on those that are widely used in taxonomic studies in biology. More detailed accounts of the area are to be found in the texts of Sneath & Sokal (1973), Jardine & Sibson (1971) and Everitt (1980); in addition a useful critique is provided in the excellent review by Cormack (1971).

First the hierarchical techniques which are particularly popular in biological studies will be introduced.

6.2 Hierarchical clustering techniques

Hierarchical clustering techniques may be subdivided into *agglomerative* methods, which proceed by a series of successive fusions of the n OTUs into groups, and *divisive* methods, which separate the set of n OTUs successively into finer groupings. In this section we shall be concerned only with the former.

Agglomerative hierarchical techniques operate on an inter-OTU proximity matrix with, initially, each OTU being considered to be a separate, single-member 'cluster'. The two OTUs having the highest similarity (or smallest distance) are then grouped together and proximities between each of the remaining OTUs and this two-member cluster calculated according to one of the methods described in Chapter 3 (see section 3.7). This process continues with, at each stage, the number of groups being reduced by one, until the final point is reached where all the OTUs are combined into a single cluster. The techniques available differ in the method they use to calculate the proximity between a single OTU and a group containing several OTUs, or between two groups of OTUs. The simplest and perhaps best known of these methods is *single-linkage clustering*; consequently, a more detailed description of this particular procedure will now be given. (Readers should note that up to now no attempt has been made to define the term *cluster*; some comments about the problems involved in achieving a satisfactory definition will be made in a later section; here the term will be used in an essentially intuitive fashion.)

6.2.1   *Single-linkage clustering*

This method defines the similarity (dissimilarity) between two clusters of OTUs as that of their most similar (least dissimilar) pair, where only pairs consisting of one OTU from each group are considered. This measure of inter-group distance is illustrated in Figure 6.1.

As an example of the operation of agglomerative hierarchical techniques in general and single linkage in particular, the method will be applied to the following dissimilarity matrix

$$
\mathbf{D}_1 =
\begin{array}{c c}
\begin{array}{c} \text{OTU} \\ 1 \\ 2 \\ 3 \\ 4 \\ 5 \end{array} &
\begin{array}{c}
\begin{array}{ccccc} 1 & 2 & 3 & 4 & 5 \end{array} \\
\left[
\begin{array}{ccccc}
0.0 & & & & \\
2.0 & 0.0 & & & \\
6.0 & 5.0 & 0.0 & & \\
10.0 & 9.0 & 4.0 & 0.0 & \\
9.0 & 8.0 & 5.0 & 3.0 & 0.0
\end{array}
\right]
\end{array}
\end{array}
$$

Figure 6.1. Single-linkage inter-group distance measure.

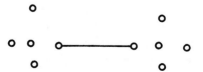

The smallest entry in the matrix is that for OTUs 1 and 2; consequently, these are joined to form a two-membered cluster, and dissimilarities between this group and the other three OTUs obtained as follows:

$$d_{(12)3} = \min\{d_{13}, d_{23}\} = d_{23} = 5.0$$
$$d_{(12)4} = \min\{d_{14}, d_{24}\} = d_{24} = 9.0$$
$$d_{(12)5} = \min\{d_{15}, d_{25}\} = d_{25} = 8.0$$

A new matrix may now be formed whose entries are inter-OTU dissimilarities and cluster-OTU values.

$$\mathbf{D}_2 = \begin{array}{c} \text{OTU} \\ (12) \\ 3 \\ 4 \\ 5 \end{array} \begin{bmatrix} 0.0 & & & \\ 5.0 & 0.0 & & \\ 9.0 & 4.0 & 0.0 & \\ 8.0 & 5.0 & 3.0 & 0.0 \end{bmatrix}$$

with columns labelled (12) 3 4 5.

The smallest entry in this matrix is that for OTUs 4 and 5 and so these are now formed into a second two-membered cluster, and a new set of dissimilarities found as follows:

$$d_{(12)3} = 5.0 \quad \text{(as before)}$$
$$d_{(12)(45)} = \min\{d_{14}, d_{15}, d_{24}, d_{25}\} = d_{25} = 8.0$$
$$d_{(45)3} = \min\{d_{34}, d_{35}\} = d_{34} = 4.0$$

These may be arranged in a matrix, $\mathbf{D}_3$:

$$\mathbf{D}_3 = \begin{array}{c} \text{OTU} \\ (12) \\ 3 \\ (45) \end{array} \begin{bmatrix} 0.0 & & \\ 5.0 & 0.0 & \\ 8.0 & 4.0 & 0.0 \end{bmatrix}$$

with columns labelled (12) 3 (45).

The smallest entry is now $d_{(45)3}$, and so OTU 3 is added to the cluster containing individuals 4 and 5. Finally the groups containing OTUs (1, 2) and (3, 4, 5) are combined into a single cluster. A useful way of displaying the results of this type of procedure is by means of a diagram such as Figure 6.2, known as a *dendrogram*.

An important point to note about the results is that the clusterings proceed hierarchically, each being obtained by the merger of clusters from the previous level. For example, if at the third stage of the procedure we had clusters (1, 2, 4) and (3, 5), we would not have had a hierarchical system since neither of these clusters are obtainable by the merger of clusters present at the preceding stage.

6.2.2   *Complete-linkage clustering*

Complete-linkage clustering is the exact antithesis of single-linkage clustering, with similarity (dissimilarity) between groups now being defined as that of the least similar (most dissimilar) pair, one from each group. Using this technique on the matrix $\mathbf{D}_1$ of the previous section, we begin again by merging OTUs 1 and 2. The dissimilarities between this group and the three remaining OTUs now become

$$d_{(12)3} = \max \{d_{13}, d_{23}\} = d_{13} = 6.0$$
$$d_{(12)4} = \max \{d_{14}, d_{24}\} = d_{14} = 10.0$$
$$d_{(12)5} = \max \{d_{15}, d_{25}\} = d_{15} = 9.0$$

The final dendrogram obtained by applying complete linkage to $\mathbf{D}_1$ is shown in Figure 6.3.

Figure 6.2. Single-linkage dendrogram.

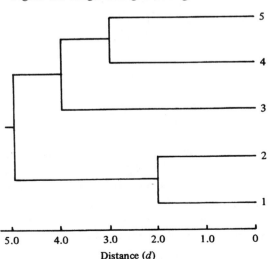

Distance (*d*)

Figure 6.3. Complete-linkage dendrogram.

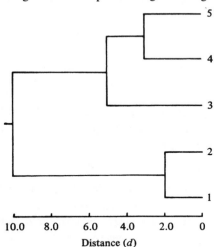

Distance (*d*)

6.2.3 *Group-average clustering*

This method defines the proximity between two clusters as the average of the proximities between all pairs of OTUs that are made up of one OTU from each group. Such a measure is illustrated in Figure 6.4.

Applying the method to the matrix $\mathbf{D}_1$ given in section 6.2.1, one begins, as with single and complete linkage, by forming a cluster from OTUs 1 and 2. A new set of dissimilarities is now defined as follows:

$$d_{(12)3} = \tfrac{1}{2}(d_{13} + d_{23}) = 5.5$$
$$d_{(12)4} = \tfrac{1}{2}(d_{14} + d_{24}) = 9.5$$
$$d_{(12)5} = \tfrac{1}{2}(d_{15} + d_{25}) = 8.5$$

Figure 6.4. Group-average inter-group distance measure.

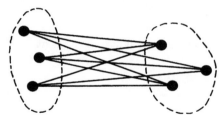

Arranging these in the matrix $\mathbf{D}_2$ we have

$$\mathbf{D}_2 = \begin{array}{c} \text{OTU} \\ (12) \\ 3 \\ 4 \\ 5 \end{array} \begin{array}{cccc} (12) & 3 & 4 & 5 \\ \left[\begin{array}{cccc} 0.0 & & & \\ 5.5 & 0.0 & & \\ 9.5 & 4.0 & 0.0 & \\ 8.5 & 5.0 & 3.0 & 0.0 \end{array}\right] \end{array}$$

The smallest entry is d_{45} and so a second cluster is formed from OTUs 4 and 5. The group average distance between the two two-membered clusters is given by

$$d_{(12)(45)} = \tfrac{1}{4}(d_{14} + d_{15} + d_{24} + d_{25}) = 9.0$$

and the procedure would continue as described in section 6.2.1.

The three methods described above operate directly on the proximity matrix and do not need access to the original character values for the OTUs. A method which does require the original data is centroid clustering.

6.2.4   *Centroid clustering*

With this method groups once formed are replaced by their mean vectors, and inter-group distance is defined as the distance between these means (see Chapter 3 and Figure 6.5). (The use of mean character values implies that we have interval scale data.)

To illustrate how this method operates it will be applied to the following set of bivariate data.

OTU	Character	
	1	2
1	1.0	1.0
2	1.0	2.0
3	6.0	3.0
4	8.0	2.0
5	8.0	0.0

Figure 6.5. Centroid inter-group distance measure.

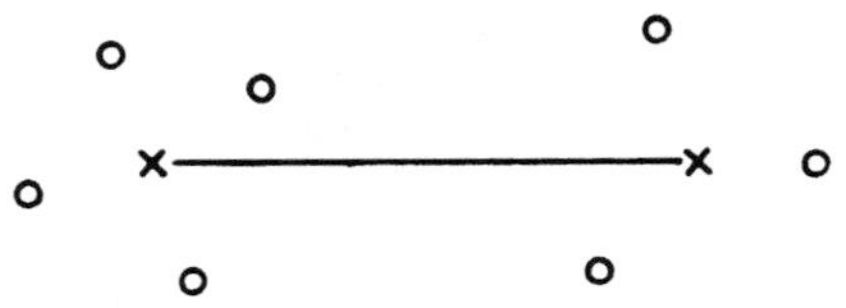

First we calculate the matrix of Euclidean distances between each pair of OTUs to give

$$
\mathbf{D}_1 =
\begin{array}{c}
\text{OTU} \\
1 \\
2 \\
3 \\
4 \\
5
\end{array}
\begin{array}{ccccc}
1 & 2 & 3 & 4 & 5 \\
\left[\begin{array}{ccccc}
0.0 & & & & \\
1.0 & 0.0 & & & \\
5.39 & 5.10 & 0.0 & & \\
7.07 & 7.00 & 2.24 & 0.0 & \\
7.07 & 7.28 & 3.61 & 2.0 & 0.0
\end{array}\right]
\end{array}
$$

The first stage of the procedure consists of fusing the two OTUs that are closest. Examination of $\mathbf{D}_1$ shows that d_{12} is the smallest entry and so OTUs 1 and 2 are fused to form a group, and the coordinates of its mean vector calculated. A distance matrix $\mathbf{D}_2$ is now computed from the following reduced data set:

OTU	Character	
	1	2
(12)	1.0	1.5
3	6.0	3.0
4	8.0	2.0
5	8.0	0.0

and we have

$$
\mathbf{D}_2 =
\begin{array}{c}
\text{OTU} \\
(12) \\
3 \\
4 \\
5
\end{array}
\begin{array}{cccc}
(12) & 3 & 4 & 5 \\
\left[\begin{array}{cccc}
0.0 & & & \\
5.22 & 0.0 & & \\
7.02 & 2.24 & 0.0 & \\
7.16 & 3.61 & 2.0 & 0.0
\end{array}\right]
\end{array}
$$

The smallest entry in $\mathbf{D}_2$ is that between OTUs 4 and 5, which are now fused to form a second group, and the OTUs replaced by the coordinates of the group mean

OTU	Character	
	1	2
(12)	1.0	1.5
3	6.0	3.0
(45)	8.0	1.0

and a new distance matrix $\mathbf{D}_3$ is computed

$$\text{OTU} \quad (12) \quad 3 \quad (45)$$

$$\mathbf{D}_3 = \begin{array}{c} (12) \\ 3 \\ (45) \end{array} \begin{bmatrix} 0.0 & & \\ 5.22 & 0.0 & \\ 7.02 & 2.83 & 0.0 \end{bmatrix}$$

Now the smallest entry is that for OTU 3 and the group comprising OTUs 4 and 5, and so these are fused to form a three-membered group. The final stage consists of the fusion of the two remaining groups into one. The dendrogram for this example appears in Figure 6.6.

Several other agglomerative hierarchical clustering techniques are available (for details, see Everitt, 1980). A point to note here is that the clusters given at any level of the dendrogram produce a *partition* of the OTUs, that is, each OTU belongs to only a *single* cluster, and that the groups together contain *all* the OTUs. In biological applications it is often the whole of the dendrogram which is of interest. In other disciplines, however, it is sometimes required to choose a 'best' set of clusters that describe the data, and the question

Figure 6.6. Centroid clustering dendrogram.

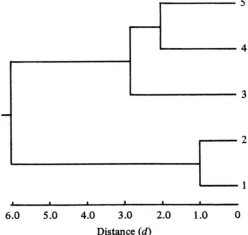

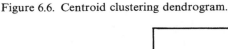

of the appropriate number of clusters becomes important. This question will be considered in section 6.4.3. Here we move on to consider briefly some of the properties of the hierarchical techniques discussed above.

6.3 Properties of hierarchical techniques

The agglomerative techniques discussed in the previous sections are *polythetic*; that is, they produce classifications based upon the complete set of recorded characters rather than on the presence or absence of a single character – a *monothetic* classification (see section 6.7 for a description of methods which produce such a classification). Both single linkage and complete linkage have the desirable property of giving results that are *invariant* under monotone transformations of the proximity matrix (cf. non-metric multidimensional scaling, Chapter 5). Consequently, when applied to different proximity matrices which have entries that are jointly monotonic (see Chapter 3), single-linkage and complete-linkage clustering will give the same results for each proximity measure. (They will *not* necessarily, however, give the same classifications as each other.) This property is not shared by group-average or by centroid clustering.

Single linkage has frequently been regarded as an undesirable clustering method because of a property known as *chaining*. This refers to the tendency of the technique to cluster together at a relatively low level in the dendrogram distinct groups of OTUs linked by a 'chain' of OTUs lying between the groups. This property is illustrated in Figure 6.7; because of it the method may fail to resolve relatively distinct groups if a number of intermediate OTUs are present.

Figure 6.7. An example of 'chaining'. The first six single-linkage fusions are indicated.

For single-linkage, complete-linkage and group-average clustering the fusion levels in the dendrogram are monotonic; that is, the fusion level at stage $i-1$ is less than that at stage i. For centroid clustering this is not necessarily so and 'reversals' of fusion levels may occur, particularly with some proximity measures, and these can be extremely troublesome (see Williams, Lambert & Lance, 1966).

In the late 1960s the first attempts at constructing a theoretical framework within which to study the properties of hierarchical techniques were made. Johnson (1967) showed that hierarchical clusters correspond to a distance measure that satisfies the *ultrametric* inequality; that is, if we consider the distance between two OTUs to be the fusion level of the dendrogram at which they first become members of the same cluster, then these distances satisfy the following inequality

$$d(x, y) \leq \max \{d(x, z), d(y, z)\} \tag{6.1}$$

Since the input similarities or distances are not generally ultrametric (and only occasionally satisfy the weaker metric inequality), Jardine & Sibson (1968) suggest that a cluster method which transforms a proximity matrix into a hierarchic dendrogram should therefore be regarded as a method whereby the ultrametric inequality is imposed on the proximity measure. They then specify a number of criteria which they argue it is reasonable for any such transformation to satisfy, and prove that single linkage is the only method satisfying them all, the implication apparently being that it is therefore the only acceptable method. This conclusion has led to a certain amount of controversy. For example, Williams, Lance, Dale & Clifford (1971) question the need for cluster methods to satisfy *all* of Jardine and Sibson's proposed criteria and adopt a more pragmatic approach to clustering, insisting that in practice single linkage did not provide solutions that investigators found useful. Again, Gower (1975) feels that Jardine and Sibson's rejection of all but single linkage is too extreme and questions whether their criteria are too stringent. He concludes that some of the criteria are unnecessary. It must be said that the approach taken by Jardine and Sibson appears to have had little impact on the majority of the users of cluster analysis; single linkage is not particularly popular and the alternative mathematically acceptable method provided by these authors is applicable only to small data sets and the solutions given are extremely difficult to interpret. (A brief description of the method is given in section 6.7.)

An alternative and very promising approach to understanding and evaluating the variety of hierarchical techniques available is to compare the effectiveness of different methods across a variety of data sets generated to have a particular structure. In this way the solutions obtained by a particular technique may be compared with the generated structure. Several studies of this type have been undertaken, for example, by Cunningham & Ogilvie (1972), Kuiper & Fisher (1975) and Blashfield (1976). In general the results of such studies indicate that

(a) no single method is best in every situation
(b) the mathematically respectable single linkage is, in most cases, the *least* successful for the data used, and
(c) group average clustering and a method due to Ward (1963) (see section 6.4.2), do fairly well, overall

Such empirical studies can, of course, never provide a complete evaluation of clustering methods, but the results obtained appear to indicate that Williams, Lance and co-workers (1971) are correct in the pragmatic approach they take and there are more *useful* clustering methods than the mathematically acceptable single linkage.

6.4   Other clustering methods

The agglomerative hierarchical techniques described earlier in this chapter are of particular interest in biological taxonomy. Nevertheless, they represent only a tiny fraction of the vast collection of clustering methods now available, and in this section a number of these other techniques will be described.

6.4.1   *Monothetic divisive clustering*

Monothetic clustering methods are generally used in cases involving binary characters. A division of the data set is then made initially into those OTUs that possess and those that lack one particular character. If only divisions of this simple type are considered then, given data for which p binary characteristics are recorded, there are p potential divisions of the initial data, $p-1$

potential divisions of each of the two clusters thus produced, and so on. The particular character chosen to generate the division is that which maximizes some dissimilarity criterion between the two groups. In general these criteria are based upon some type of chi-squared statistic; details are given in Lambert & Williams (1962, 1966), MacNaughton-Smith (1965) and Everitt (1980).

6.4.2 *Minimization of trace* (**W**)

A method of cluster analysis which is very widely used is one that attempts to find the partition of the OTUs into a specified number of groups, say k, which minimizes the sum of the within-cluster sum of squares of each character. That is, it seeks the partition of the n OTUs into k groups which has the lowest value of trace (**W**), where **W** is the $(p \times p)$ matrix obtained by summing the within-cluster sum of squares and product matrices over all k clusters; that is,

$$\mathbf{W} = \mathbf{W}_1 + \mathbf{W}_2 + \ldots + \mathbf{W}_k \qquad (6.2)$$

In theory, finding this partition is straightforward; we simply consider the value of trace (**W**) for *every* partition of the OTUs into k groups. In practice, of course, problems arise because of the enormous number of possible partitions; for example, for $n = 19$ and $k = 8$, there are 1,709,751,000,480 possible partitions. Consequently, complete enumeration is out of the question even with the fastest computers, and *hill-climbing* algorithms are used which seek to obtain a k-group partition with a lower value of trace (**W**) than an existing partition, by reassignment of OTUs to clusters. By iterating from some given starting partition, the algorithm seeks successively improved solutions until some convergence criterion is satisfied. Such a procedure cannot guarantee finding the *global* minimum of trace (**W**), only a *local* minimum (see Everitt, 1980, for more details).

With this technique solutions are usually obtained for a variety of values of k, and some method is adopted for deciding on the 'best' or most appropriate value (see, for example, Englemann & Hartigan, 1969; and Hartigan, 1975). The solutions obtained by this method are not necessarily hierarchical, but a hierarchical procedure based upon the trace (**W**) criterion is available (see Ward, 1963).

6.4.3. *A multivariate mixture model for cluster analysis*

A statistical model of clustering which takes account of variation amongst the OTUs within the same cluster has been suggested by Wolfe (1970); the assumption is made that the within-cluster distribution of character values is Gaussian with a particular vector of mean values and a particular variance–covariance matrix. Such an assumption implies that the distribution of character values in a population composed of, say, k such clusters is given by $f(\mathbf{x})$ where

$$f(\mathbf{x}) = \sum_{i=1}^{k} \lambda_i \alpha_i(\mathbf{x};\boldsymbol{\mu}_i,\boldsymbol{\Sigma}_i) \tag{6.3}$$

where $\mathbf{x}$ is a p-dimensional random variable (a realization of which is the p character values for an individual), the $\lambda_i, i = 1,\ldots,k$ are the proportions of each of the k clusters in the population and are subject to the constraints $0 < \lambda_i < 1$ and $\sum_{i=1}^{k} \lambda_i = 1$; the functions $\alpha_i(\mathbf{x};\boldsymbol{\mu}_i,\boldsymbol{\Sigma}_i)$ represent multivariate normal densities with mean vector, $\boldsymbol{\mu}_i$, and variance–covariance matrix, $\boldsymbol{\Sigma}_i$, that is,

$$\alpha_i(\mathbf{x};\boldsymbol{\mu}_i,\boldsymbol{\Sigma}_i) = (2\pi)^{-p/2}|\boldsymbol{\Sigma}_i|^{-\frac{1}{2}}$$

$$\times \exp -\tfrac{1}{2}(\mathbf{x} - \boldsymbol{\mu}_i)'\boldsymbol{\Sigma}_i^{-1}(\mathbf{x} - \boldsymbol{\mu}_i) \tag{6.4}$$

From our sample of n OTUs supposedly taken from a population described by $f(\mathbf{x})$, it is possible to estimate the parameters $\lambda_i, \boldsymbol{\mu}_i$ and $\boldsymbol{\Sigma}_i$ for $i = 1,\ldots,k$ by maximum likelihood methods; subsequently, OTUs may be associated with the particular cluster to which they have greatest posterior probability of belonging. This probability, $P(s|\mathbf{x}_i)$, is estimated by

$$P(s|\mathbf{x}_i) = \frac{\hat{\lambda}_s \alpha_s(\mathbf{x}_i;\hat{\boldsymbol{\mu}}_s,\hat{\boldsymbol{\Sigma}}_s)}{f(\mathbf{x}_i)} \tag{6.5}$$

More details of this method are given in Everitt (1980) and Everitt & Hand (1981).

6.4.4 *Jardine and Sibson's K-dend clustering method*

Sibson (1970) shows that axioms of stability, optimal cluster preservation and invariance under relabelling or any monotonic transformation of the proximity matrix lead uniquely to a system,

Table 6.1. *Hypothetical similarity matrix*

1				
2	0.3			
3	0.4	0.1		
4	0.1	0.2	0.2	
5	0.2	0.3	0.4	0.1

Figure 6.8. Illustration of the formation of Jardine & Sibson overlapping clusters. (*a*) Maximal complete subgraphs for different values of *H*. (*b*) Cluster formation for $K = 1$. (These are single-linkage clusters.) (*c*) Cluster formation for $K = 2$. (Clusters may overlap to the extent of one OTU.) (*d*) Cluster formation for $K = 3$. (Clusters may overlap to the extent of two OTUs.) (Taken with permission from Jardine and Sibson, 1968.)

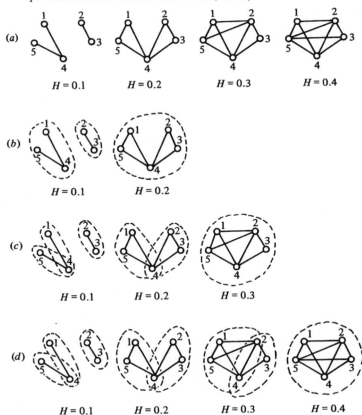

first described by Jardine & Sibson (1968), which generates a series of overlapping clusters. Essentially this method consists of representing each OTU by a node on a graph and connecting all pairs of nodes which correspond to OTUs having a similarity value above some specified threshold value, H. Next, a search is made for the largest subsets of OTUs for which all pairs of nodes are connected (these are known as *maximal complete subgraphs*). Now all pairs of maximal complete subgraphs which intersect in at least a particular number of nodes, say K, are further connected. When no more connections can be found the Jardine–Sibson overlapping classification for this value of H and K has been obtained. When $K = 1$ no overlaps occur and the procedure becomes single-linkage clustering; in general the clusters found by this method may overlap to the extent of having $K-1$ points in common. The procedure is illustrated in Figure 6.8 for the similarity matrix shown in Table 6.1. An algorithm given by Jardine & Sibson (1968) for implementing the method has been considerably improved by Cole & Wishart (1970).

6.5   **An example**

Consider Table 6.2. This contains a matrix of distances (presumed to be Euclidean) for eleven forms of the bee *Hoplites*

Table 6.2. *Matrix of distance coefficient (based on standardized data) for the forms of the* Hoplites producta *complex (Michener, 1970)*

	1	2	3	4	5	6	7	8	9	10	11
1	0										
2	0.940	0									
3	1.229	0.791	0								
4	1.266	0.847	0.303	0							
5	1.507	1.331	1.070	1.026	0						
6	1.609	1.306	0.778	0.573	1.175	0					
7	1.450	1.266	1.475	1.506	1.829	1.876	0				
8	1.239	1.286	1.510	1.540	1.908	1.832	1.655	0			
9	1.493	1.160	0.848	0.792	0.965	0.978	1.847	1.761	0		
10	1.494	1.396	1.497	1.528	1.724	1.687	1.954	1.733	1.721	0	
11	1.348	1.238	1.352	1.385	1.724	1.559	1.844	1.608	1.596	0.645	0

The names of the forms of *Hoplites* are 1, *Hoplites gracilis*; 2, *subgracilis*; 3, *interior*; 4, *bernardina*; 5, *panamintana*; 6, *producta*; 7, *colei*; 8, *elongata*; 9, *uvularis*; 10, *grinelli*; 11, *septentrionalis*.

producta, so similar that a total of only 23 characters were found to vary among the members of the group (Michener, 1970). Since this distance matrix is rather difficult to interpret as it stands, one needs a way or ways of summarizing the data. First consider the results of two different methods of ordination. Figure 6.9(*a*) gives a three-dimensional representation of the character space, produced by the method of principal co-ordinates analysis (Michener, 1970). Broken lines show the minimum spanning tree for these taxa. Following the broken lines, and using a distance of over 1.0 as a criterion of a gap

Figure 6.9. (*a*) Plot of first three principal coordinates for eleven species of *Hoplites* with superimposed minimum spanning tree. (Taken with permission from Michener, 1970.)
(*b*) Multidimensional scaling of the same eleven species of *Hoplites*.

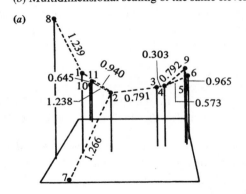

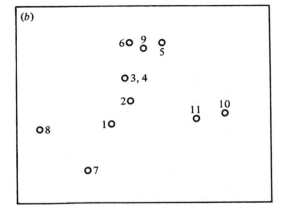

or discontinuity, one comes to the following conclusions. OTUs 1 to 6 and 9 form a single elongate cluster, OTU 7 and OTU 8 seem to be well isolated from the rest of the group and from each other and, finally, OTUs 10 and 11 form an isolated cluster. Figure 6.9(*b*) shows a two-dimensional representation of the same data, produced by non-metric multidimensional scaling (analysis by the present authors). The configuration of the eleven points corresponds, more or less, to that produced by principal coordinates analysis, and the conclusions that one can draw from the results are the same.

Now consider Figure 6.10(*a*). This is a dendrogram, based on the distances given in Table 6.1, produced by group-average clustering (see section 6.2.3). If one produces a partition of the eleven OTUs, by drawing a vertical line through this dendrogram at a distance of about 1.4, it is clear that the resulting classification is the same as

Figure 6.10. (*a*) Group average clustering of eleven species of *Hoplites* using Euclidean distances.
(*b*) Group average clustering of eleven species of *Hoplites* using correlation coefficient. (Taken with permission from Michener, 1970.)

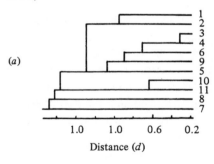

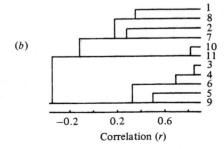

that produced from the ordinations. However, it is also clear that OTUs 1 and 2 form a distinct subcluster of the larger cluster. Examination of the original distance matrix does not suggest that OTUs 1 and 2 are, for example, isolated from OTU 3, and neither does either of the ordinations. The dendrogram appears to have produced a subcluster that is an artefact of the analysis. To illustrate how easy such changes are to produce, Figure 6.10(b) shows a dendrogram produced by the same clustering algorithm, but in this case based on the use of correlations as measures of proximity of the same eleven OTUs (Michener, 1970). Similar changes could have been produced quite readily by changing the clustering algorithm rather than the measure of proximity (see, for example, Moss, 1967). It must be stressed to the reader that, with the ready availability of clustering and ordination algorithms, it is the evaluation of the results (along with the initial choice of characters and a proximity measure) that is vital for the success of a taxonomic exercise that involves the use of numerical methods. Some suggestions that may be useful in such evaluation are discussed in the following section.

6.6   The evaluation of results and other problems

Since clustering techniques will generate a set of clusters even when applied to random, unclustered data, the question of validating and evaluating solutions becomes of great importance. Essentially this means that we wish to address the question: are the clusters and structures generated by a clustering method or algorithm 'significant' enough to provide evidence for hypotheses about the phenomena being studied? Now in some cases, for example when one has a great deal of experience with a particular clustering method and some prior information about the data being clustered, interpreting the results of clustering methods becomes a personal matter in which intuition and insight are dominant. However, the user of a clustering algorithm is often unsure about the data and has little experience with a particular type of data or a particular clustering method. Indeed lack of information about the data is often the motivation for clustering the data in the first place, and in such cases the user searches for objective meaning and needs quantitative measures of significance for evaluating clustering procedures. A considerable amount of work has been undertaken in this area, and an excellent review is provided by Dubes & Jain (1979). The proposed methods may be categorized by the questions they attempt to answer, although the questions are clearly not mutually exclusive.

(a) Is the data matrix random? Unless some evidence exists that the data tend to cluster, there is little basis for imposing any cluster structure on it.

(b) How well does a hierarchy fit a proximity matrix? A high degree of global fit between dendrogram and proximity matrix is necessary if all the clusters are to be meaningful.

(c) Is a partition obtained from one level of a hierarchy valid?

(d) Which individual clusters appearing in a hierarchy are valid?

Let us now briefly consider the methods which have been proposed to deal with each of these questions. For a more detailed account the reader is again referred to the paper by Dubes & Jain (1979).

6.6.1 *Measuring clustering tendency*

Attempts to deal with the first of the four questions above have centred on deciding on an appropriate null hypothesis of 'no clustering'. Two possibilities which have been considered are the *random graph hypothesis* and the *random pattern hypothesis*. The first is applicable in studies involving symmetric proximity matrices whose entries are rank orderings; that is, the most similar pair of OTUs have rank one, and the least similar rank $n(n-1)/2$. The random graph hypothesis is that all $[n(n-1)/2]!$ such matrices are equally likely. The random position hypothesis views the np-dimensional observations as independent samples from some p-dimensional distribution which would imply lack of any cluster structure, such as uniform, or unimodal Gaussian.

Ling (1972) and Ling & Killough (1976) have considered the random graph hypothesis and produced suitable probability distributions; details, however, are outside the scope of this text. Strauss (1975) and Saunders & Funk (1977) have considered the random position hypothesis and derive a statistic based on the number of inter-point distances that are less than some threshold value; under the hypothesis of a uniform distribution for the data, this statistic can be shown to have a Poisson distribution.

6.6.2   *Global fit of hierarchy*

Hierarchical clustering techniques impose a hierarchical structure on data and we need to consider whether this is merited or whether it introduces unacceptable distortions of the original relationships between the OTUs as implied by their observed proximities. The most common method for evaluating the match between the dendrogram and the proximity matrix is the *cophenetic correlation coefficient*. This is simply the product moment correlations of the $n(n-1)/2$ entries in the lower half of the measured proximity matrix, with the $n(n-1)/2$ entries in the *cophenetic* matrix $\mathbf{C}$, where c_{ij} is defined as the first level in the dendrogram at which OTUs i and j occur in the same cluster. Since these values satisfy the ultrametric inequality (see section 6.3), the match between data and dendrogram cannot be perfect unless the entries in the proximity matrix are also ultrametric, a situation which seldom occurs in practice.

To illustrate the use of the cophenetic correlation coefficient we may use the data given in section 6.2.1. The elements of $\mathbf{D}$ and $\mathbf{C}$ to be correlated are as follows:

$$d_{ij}: 2.0 \quad 6.0 \quad 5.0 \quad 10.0 \quad 9.0 \quad 4.0 \quad 9.0 \quad 8.0 \quad 5.0 \quad 3.0$$
$$c_{ij}: 2.0 \quad 5.0 \quad 5.0 \quad 5.0 \quad 5.0 \quad 4.0 \quad 5.0 \quad 5.0 \quad 4.0 \quad 3.0$$

The cophenetic correlation takes the value 0.82. Rohlf & Fisher (1968) studied the distribution of the cophenetic correlation under the hypothesis that the OTUs are randomly chosen from a single multivariate Gaussian distribution. They found that the average value of the coefficient tended to decrease with n and to be almost independent of the number of characters recorded, and that a value above 0.8 was usually sufficient to reject the null hypothesis. However, in a later paper, Rohlf (1970) warns that 'even a cophenetic correlation near 0.9 does not guarantee that the dendrogram serves as a sufficiently good summary of the phenetic relationships'.

6.6.3   *Partitions from a hierarchy*

In hierarchical clustering, partitions are achieved by cutting a dendrogram or selecting one of the solutions in the nested sequence of clusterings that comprise the hierarchy. In particular applications it may be of interest to try and determine which of all the possible partitions provides the best fit to the data; essentially, this means that we have to decide on the appropriate number of clusters for the data. One informal method which is often used for this purpose is to examine the difference between fusion levels in the dendrogram. Large differences are taken to imply a particular number of clusters. For example, consider the dendrogram shown in Figure 6.11. This shows a large difference in the level between two groups and the final stage at which all OTUs are in a single group. This would be taken as evidence for the two-group solution.

A more formal approach to this problem is described by Mojena (1977), who describes two possible 'stopping rules'. From empirical studies described in the paper, one of these rules does appear worthy of further consideration as a pragmatic means of objectively assessing the selection of a particular partition from a hierarchical clustering.

Apart from techniques designed to deal with the specific questions outlined above much can be learnt about solutions informally with various graphical displays, for example the result of an ordination technique and superimposed minimum spanning tree as discussed in the previous chapter. A number of other graphical methods are available which can be useful in particular cases. One of these involves simple displays of various distances, which can be useful for studying tightness of and separations amongst the clusters. For example, suppose there are only two variables and five OTUs, and the application of a clustering algorithm has produced three clusters, A, B and C. In Figure 6.12(a) the distance of each OTU to each cluster mean is plotted. For example, the first column shows the distances of all five OTUs (labelled by their cluster membership as determined by the clustering algorithm) to the mean of cluster A. The focus in this plot is on the isolation and tightness of each cluster. Cluster A, for instance, is isolated from the other two and is a fairly 'tight' cluster. In Figure 6.12(b) the same distances are plotted in a different way. Now the focus is on OTUs and the strength of their classification into clusters; this plot shows how far away each cluster centroid is from each OTU. For example, although OTU A_1 clearly belongs to cluster A, OTU B_2 which belongs to cluster B is also relatively close to cluster C.

Some other simple plots of distances which can be useful are described in Gnanadesikan, Kettenring & Landwehr (1977).

Figure 6.11. Dendrogram indicating the presence of two distinct clusters.

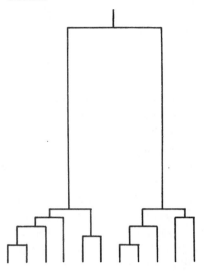

Figure 6.12. Plots of OTU-to-cluster distances. A_1 and A_2 from cluster A, B_1 and B_2 cluster B, and C is a single-member cluster. (Taken with permission from Gnanadesikan, Kettenring and Landwehr, 1977.)

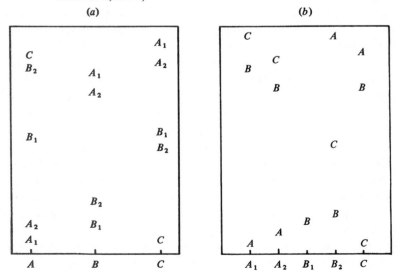

A further graphical aid which is sometimes useful in interpreting the results of a cluster analysis is that originally proposed by Andrews (1972). The essential idea is that a p-dimensional observation may be mapped into a function of the following form:

$$f_x(t) = x_1/\sqrt{2} + x_2 \sin t + x_3 \cos t + x_4 \sin 2t + x_5 \cos 2t + \ldots$$

and then this function is plotted for $-\pi < t < \pi$. Each observation is thus represented as a line on such a plot. The particular function used is especially useful since points that are close together in the original p dimensions (as judged by their Euclidean distance) will be represented by lines that remain close together for all values of t. Consequently, the plots might be useful for identifying clusters of observations from the original data, or for displaying the results of a cluster analysis by plotting cluster means. Figure 6.13 shows the Andrews plots of 30 five-dimensional observations. The figure clearly indicates the presence of three clusters of observations.

In addition to the difficulties involved in interpreting solutions, the taxonomist also faces other problems. If different measures of proximity produce different classifications when, for example, using a common method of clustering, which proximity measure should one choose? A similar choice has to be made with respect to the clustering algorithm to be used. Different sets of characters, or samples of OTUs, will inevitably produce data that indicate different classifications, although they might be reasonably close. For example, if one were to classify plants on the basis of vegetative characteristics one might produce a classification differing quite widely from one produced from measurements of seed characters. Does this matter? Michener (1970) has suggested that

> it is increasingly evident that a major value of numerical phenetics is the possibility of preparing a variety of classifications using different methods or different sets of characters for comparative purposes. For example, it may be very useful to prepare separate classifications for larvae and for adults. The classification most useful for predictive purposes for larval characters will probably be that based upon other larval characters... Once the data are collected and coded for numerical phenetic use, additional manipulation of them

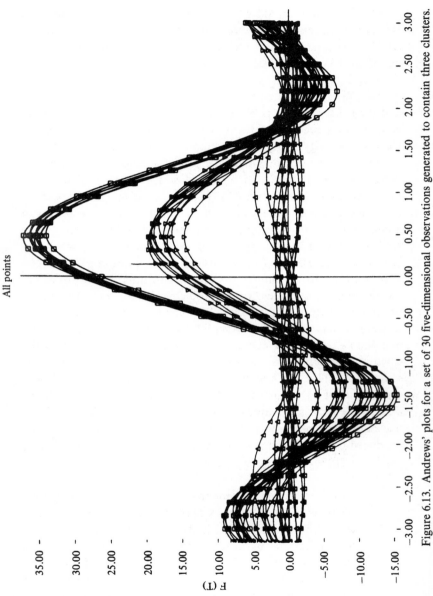

Figure 6.13. Andrews' plots for a set of 30 five-dimensional observations generated to contain three clusters.

can be made so rapidly by computer that there is little reason for not trying out all possibly interesting procedures.

One might wish, however, to corroborate the results of one study by examination of a second character set. If, for example, a dendrogram or scatter plot has been produced from an analysis of morphological character states, one might wish to test the usefulness of the results by repeating the work on biochemical characters such as amino acid sequences or antigenic similarities. In order to assess the results one would clearly need some quantitative measure of congruence between classifications. However, this is beyond the scope of the present text and the interested reader is referred to Sneath & Sokal (1973) or Jardine & Sibson (1971) for a fuller discussion.

6.7   What is a cluster?

Throughout the earlier parts of this chapter the terms cluster and group have been used in an essentially intuitive manner, without any attempt at formal definitions. However, this problem needs to be tackled at this point in order to clarify the way in which the different cluster analysis algorithms have been developed to search for particular types of grouping. When a set of OTUs have been partitioned or classified in some way the taxonomist wishes to be able to describe how the members of a group are similar, and how these members differ from those of other groups. For example, when considering what is meant by a biological species, a zoologist such as Mayr (1969) uses the ability to interbreed as a criterion of membership. Reproductive barriers are seen as the natural 'gaps' between species. However, numerical taxonomists have rejected the biological species concept (see section 1.4) and therefore need other ideas with which to work. Here the obvious choice is to use measurements of similarity or distance to indicate relationships within and between groups.

Consider Figure 6.14 which is a representation of the positions of 60 hypothetical OTUs in a two-dimensional character space. It is intuitively clear that there are two distinct clusters, but how could one produce a clustering algorithm to detect them? Since an algorithm usually follows from a precise definition of what is required, this is a problem of cluster definition. First note that the clusters are of different size (in terms of space covered) and shape, and, although many clustering algorithms are written on the assumption that clusters might be of similar size and shape, there is no *a priori* reason why this should be so. Now, if one inspects group *A*, and assumes that the Euclidean distance between OTUs is a sensible measure of proximity or relatedness, it is clear that all members of this group are within 0.8 of any of the others. In terms of graph theory (see, for example, Ore, 1963), if one defines two OTUs to be *connected* (or related) if the distance between them is less than or equal to 0.8, group *A* is clearly represented by a *maximally connected graph*; that is, there are direct connections between every pair of

Figures 6.14. Positions of 60 hypothetical OTUs in a two-dimensional character space.

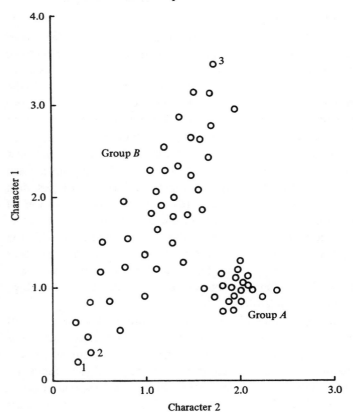

OTUs. This leads naturally to one possible definition of a cluster as a maximally connected set. One produces these sets using the complete linkage clustering algorithm. Now, if one also uses the same criterion for connectivity for group *B*, these OTUs form a graph that is *not* maximally connected. In order for this to appear as a maximally connected set one would have to redefine what is meant by 'connected'. If one now defines two OTUs to be connected if the distance between them is less than or equal to 3.6 (the approximate distance between OTUs 1 and 3), group *B* will form a maximally connected set. But so will all the OTUs taken together! If one were to define connectivity by an intermediate distance artefactual clusters would be produced by the complete-linkage clustering method. Redefining a cluster as a group of OTUs that were *minimally connected*, that is, by only one direct or indirect path between every pair of OTUs, and searching for them using the single-linkage clustering algorithm, one might recover the true structure of the data. However, the possibility of chaining (see section 6.3) might again lead one astray. Perhaps one might wish to define a cluster to be intermediate between a maximally and minimally connected set in an attempt to avoid the above problems.

It should be clear from the above discussion that no one definition of a cluster will be suitable for all sets of data (this is an alternative way of expressing the conclusions of section 6.3), and that the definition chosen should be dependent on the results of, for example, ordination. Instead of thinking in terms of graph theory, one might wish to consider areas of relatively high density of OTUs as clusters. One might also wish to consider the idea of a *gap* or *moat* that surrounds a distinct cluster. This is, perhaps, what one is intuitively looking for when visually examining scatter plots such as Figure 6.14. Michener (1970) has suggested that more importance should be given to these discontinuities by phenetic taxonomists, since these gaps are often of great biological significance due to the historical and ecological factors that have led to them.

Having decided to use a particular clustering algorithm and having produced a dendrogram to summarize the results, one needs to decide on appropriate points at which to cut the dendrogram to produce sensible partitions of the OTUs into taxa (called *phenons* by Sneath & Sokal, 1973). Then, on the assumption that one wishes to produce a classification comparable to that of traditional taxonomists, one has to decide on the *rank* of the taxa. These are problems that are not unique to numerical taxonomy, but it might be argued by numerical taxonomists that the traditional taxonomies should not be used as a precedent when deciding how to summarize information of this type. It has also been argued that the traditional type of binomial nomenclature is of little use to the modern systematist. If one merely uses the results of ordination and cluster analysis as summaries of the data, one can then use one's experience and insight as a taxonomist to produce a formal classification that will best serve the purposes of biologists. It was stated in Chapter 1, however, that one of the principal aims of numerical taxonomy was to produce classifications on the basis of objective criteria in the hope that they would be stable and in some way would be 'better' than those produced by traditional taxonomists. One cannot escape from the conclusion that skill and insight are just as important to the numerical taxonomist as they are to a worker who does not use mathematical methods. This view is succinctly summarized by Michener (1970):

> It is clear...that numerical phenetics has not succeeded in developing a method of producing a single relatively objective classification, better than others, which could be used as the formal classification of any group.
> ...The selection of characters, the coefficients of similarity or difference to be used, and the methods of clustering are all determined subjectively, and there are no generally accepted criteria for deciding which of various classifications is 'best'.

6.8   **Summary**

A large number of algorithms for cluster analysis have been developed and many are available in the form of well-documented computer packages. Consequently they are easy to use. They are also, however, easy to misuse and to misinterpret. Both the results of cluster analyses and those of ordinations have to be examined carefully by a skilled taxonomist. These methods, in practice, do not lead to a purely objective and stable classification of OTUs. We agree with Michener (1970) in concluding that these methods are best seen as tools for data exploration, rather than for the production of a formal classification. With this in mind, it would appear that the optimism of numerical taxonomists such as, for example, Sokal & Rohlf (1970) is not really justified. These conclusions, however, are not to be interpreted as criticisms of numerical methods, but are merely intended to imply that one cannot *replace* careful thought by automatic computerized methods.

7

Identification and assignment techniques

7.1   Introduction

Once one has a comprehensive classification of the taxonomic units of interest, the next problem becomes that of assigning a new unit to a predetermined category or taxon. How does one identify a freshly collected specimen? Determination or identification by an expert is perhaps still regarded as the most reliable of all identification methods. Here the scientist will use experience, intuition and skill in comparing the new organism with examples, illustrations and descriptions of the previously constructed taxa. Often a specimen will be immediately recognized by the experienced scientist without any specific mental procedure. One cannot always have recourse to an expert, however, and it therefore becomes necessary to consider designing methods of identification that might be used by a beginner and, ideally perhaps, programmed for a computer. By supplying a precise algorithm for identification decisions will be made objectively, with the same result arising on repeated consideration of the same evidence.

There are two main approaches to the automatic classification of specimens. In the first one employs characters in a *sequence* (as in a *diagnostic key*). Here the possible alternatives are successively eliminated by considering more and more characters until only one possibility remains. In the second approach one considers all of the chosen characters *simultaneously*. Here some sort of matching is made between the unknown specimen and each of the known taxa in order to find the best match. As with that of cluster analysis, the literature for identification or diagnosis is large and widely dispersed and, consequently, it will not be possible to describe every available method in a single chapter. Instead this chapter will begin with an introduction to the construction of diagnostic keys, and will then concentrate on the more mathematical matching techniques such as *discriminant function* and *canonical variate analysis,* which are being used increasingly in biological applications.

7.2 **Diagnostic keys**

These keys are perhaps the most familiar of identification methods. A simple example in which the taxa are eleven species of common British tree is shown in Table 7.1, taken from Payne & Preece (1980). This key consists of seven numbered sets of contrasting statements, each set relating to the states of a single character. Each of the statements is a *lead* describing a single character state. In *binary* or *dichotomous* keys there are only two leads per set of statements, and these sets are referred to as *couplets.* To identify a specimen from one of the eleven taxa one starts by choosing from the first set of leads the one that is true for the specimen. The chosen lead may be followed by a name identifying the appropriate taxon, or by a number to the right of the lead directing one to a further set where a decision is made as before, the process continuing until an identification is made. In summary, the use of a diagnostic key involves performing a sequence of tests, each of which has a number

Table 7.1. *Key to eleven types of common British tree (Payne & Preece,* 1980)

1. Texture of bark smooth	2
Texture of bark rough	4
Texture of bark corky	Elder
Texture of bark scored horizontally	Rowan
Texture of bark scaling	6
2. Leaves not pinnate or lobed	3
Leaves pinnate	Ash
3. Basic shape of leaf pointed oval	Beech
Basic shape of leaf heart-shaped	Lime
4. Leaves not pinnate or lobed	5
Leaves lobed	Oak
5. Basic shape of leaf pointed oval	Elm
Basic shape of leaf broad lanceolate	Sweet chestnut
6. Leaves not pinnate or lobed	Birch
Leaves lobed	7
7. Position of leaves on stem opposite	Sycamore
Position of leaves on stem alternate	Plane

of different possible outcomes or responses. For each of the tests, the response matching the character state of the unknown specimen is selected.

Table 7.1 can be represented diagrammatically by a tree as shown in Figure 7.1. For such diagrams it is natural to define a *branch* of a key as a sequence of tests and responses that result in the assignment of the unknown specimen to a given taxon, and to consider each test to be made at a particular *point* in the key. The *depth* of a test is the number of tests preceding it on a branch, and the *length* of a branch is similarly the total number of tests leading to a given identification.

The characters best suited for construction of a diagnostic key are those that are both easy to observe and convey as much information as possible about the differences between taxa. They are characters that have the highest consistency within taxa, but separate the set of taxa under consideration into approximately equal halves. There is always a possibility of making one or more mistakes when using a diagnostic key; the earlier the mistake is made in the sequence the more serious the consequences. When constructing a key using a given set of characters one therefore wishes to reduce the probability

Figure 7.1. Diagrammatic representation of the diagnostic key shown in Table 7.1. Open circles indicate questions and crosses indicate taxa. (Taken with permission from Payne and Preece, 1980.)

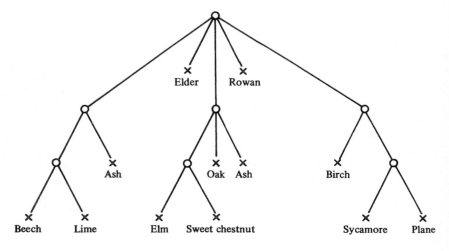

of making a mistake, and this is done by keeping the mean branch length of the key as small as possible (although other considerations may override this – see Pankhurst, (1978)). Figure 7.2 shows two extreme cases of diagnostic keys to distinguish eight hypothetical taxa. For key *a* the mean branch length is

$$\frac{3 + 3 + 3 + 3 + 3 + 3 + 3 + 3}{8} = 3.0$$

For key *b* it is

$$\frac{1 + 2 + 3 + 4 + 5 + 6 + 7 + 7}{8} = 4.4$$

The difference between the shortest and longest key is more marked if many more taxa are being discriminated. For example, with 64 taxa the shortest and longest length keys have mean branch lengths of 6 and 32.5 respectively.

Payne & Preece (1977) describe a method for incorporating checks

Figure 7.2. Two diagnostic keys for eight hypothetical taxa. (*a*) minimum mean branch length. (*b*) maximum mean branch length. Symbols as in Figure 7.1.

(*a*)

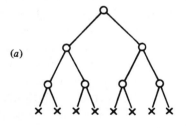

(*b*)

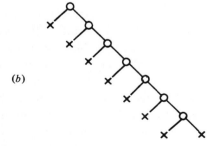

against observer error into identification keys. This involves the assumption that the use of the key will never make more than n errors and eliminates taxa from a branch only after *more* than n test results have differed from those on the branch. An *initial* or *main key* is constructed in the usual way and then, for each of its endpoints, a *check key* is constructed to distinguish between the identified taxon and taxa to which the unknown might have been assigned but for the occurrence of up to n errors. In a check key, taxa are eliminated only after $n + 1$ test results that differ either in the main key or in the check key itself. Tests from the main key are re-used if their results cannot be verified independently. Other methods for dealing with errors in diagnostic keys are described by Morse (1971) and Sandvik (1976).

7.2.1   *The construction of diagnostic keys*

Keys have long been constructed by hand, and a description of the methods used is given by Pankhurst (1978). With the increasing use of computers, however, more attention has been focussed on finding *optimal* identification keys which have, on average, as few tests per identification as possible (i.e. a minimum mean branch length; see above). The logical basis for construction of such optimal keys is, essentially, the mathematical *Théorie des Questionnaires* expounded by Picard (1965), an English version of which has recently appeared (Picard, 1980). Much of this work is outside the scope of this text and we shall therefore confine our attention to a brief account of computer algorithms for constructing keys.

Except for dynamic programming algorithms, which effectively enumerate all possible keys (Garey, 1972), no exact algorithm is known for finding optimum keys. Dynamic programming algorithms, however, are impracticable for most real data, which may be concerned with several hundred taxa and, in some cases, of the order of a hundred characters or more. Several authors, for example, Pankhurst (1970), Morse (1971), and Payne (1975) present algorithms giving approximate solutions. They all operate by selecting first the test that 'best' divides the taxa into two sets. Various criteria, some of which are described below, have been used to define what is meant by the best test. After the first division, the chosen criterion is used to select the next test to be used with each subset of taxa, and so on. Garey & Graham (1974) give examples showing that selecting tests in this way, without examining their later consequences, can lead to inefficient keys, but most authors claim that their algorithms work well in practice and certainly give keys as good as, if not better than, those prepared by hand using intuition and experience.

For tests which have equal costs and taxa for which there are no variable responses, the most common criterion used to choose the best test is based on the entropy function of Shannon (1948) and is given by

$$H_i = \sum_{k=1}^{m_i} p_{ik} \log p_{ik}, \tag{7.1}$$

where p_{ik} is the proportion of taxa with fixed response k to test i and m_i is the number of levels of test i. At each stage the test with minimum value of H_i is chosen.

For taxa having variable responses, Shwayder (1971, 1974) suggested that H_i be modified to

$$H_i' = H_i - (1 - r_i) \log (1 - r_i), \tag{7.2}$$

where r_i is the proportion of taxa with variable responses to test i.

A further function, suggested by Brown (1977), is

$$M_i = - \sum_{k=1}^{m_i} p_{ik}(1 - p_{ik} - r_i) \tag{7.3}$$

Gower & Payne (1975) investigated the properties of several such criteria and in the multiresponse case found the criterion

$$S_i = \sum_{k=1}^{m_i} (p_{ik} + r_i) \log (p_{ik} + r_i) \tag{7.4}$$

to be most suitable.

As an example of the use of these criteria, let us consider H_i' applied to a hypothetical data set consisting of five taxa and five characters (Table 7.2). For these data H_i' takes the values

$$H_1' = 0.4 \log 0.4 + 0.6 \log 0.6 = -0.29$$

$$H_2' = 0.6 \log 0.6 + 0.2 \log 0.2 - 0.8 \log 0.8 = 0.90$$

$$H_3' = 0.6 \log 0.6 - 0.6 \log 0.6 = 0.0$$

$$H_4' = 0.4 \log 0.4 + 0.2 \log 0.2 + 0.2 \log 0.2 - 0.8 \log 0.8$$
$$= -0.36$$

$$H_5' = 0.8 \log 0.8 - 0.8 \log 0.8 = 0.0$$

Table 7.2. *Characteristics of five hypothetical taxa*

Character	Taxa				
	T_1	T_2	T_3	T_4	T_5
1	1	1	2	2	2
2	V	2	3	2	2
3	2	V	V	2	2
4	V	1	1	2	3
5	2	2	2	2	V

The numbers in the table indicate the character states diagnostic for particular taxa. 'V' indicates a variable response.

Consequently, the first character chosen is character 4; this leads directly to identification as T_4 for character state 2, and T_5 for character state 3; a further character must be chosen to differentiate between T_1, T_2 and T_3. By applying the same procedure this is found to be character 2. For these data the remaining characters are unnecessary for identification.

7.3   Probabilistic assignment techniques

If identification with certainty is impossible, either because too many characters are variable within taxa or because all assessments of character states are subject to error, probabilistic identification methods are often used. The question now asked is: which taxon is *most likely* to have produced the character states of the specimen requiring identification?

First consider a single character found to have state a in an unknown specimen S. What one wishes to determine is the probability of S belonging to each of the taxa under consideration. The specimen may then be assigned to the taxon corresponding to the maximum of these probabilities. By *Bayes' theorem* (see, for example, Hays, 1973), the probability of a specimen S with character value a, belonging to taxon T_i, designated by $P(T_i|a)$, may be expressed as follows:

$$P(T_i|a) = \frac{P(a|T_i)P(T_i)}{P(a)} \tag{7.5}$$

where $P(a|T_i)$ is the probability of observing character state a in taxon T_i, $P(T_i)$ is the prior probability of finding a specimen belonging to taxon T_i, and $P(a)$ is the probability of observing character state a; the latter may be written in terms of conditional probabilities of observing character state a in each of the n taxa involved, as follows:

$$P(a) = \sum_{i=1}^{n} P(a|T_i)P(T_i) \tag{7.6}$$

Both (7.5) and (7.6) still hold if we consider a to be a vector (**a**) of character states observed on S. The form of $P(T_i|\mathbf{a})$ corresponding to (7.5) shows that the probability of S belonging to T_i is high if **a** is a common character set for that taxon, and also high if T_i is a commonly occurring taxon (the corollary of this, of course, is that the probability of being in taxon T_i given **a**, is low if that taxon is a very rare one).

One of the assumptions commonly used for qualitative character states is that the p characters vary independently so that

$$P(\mathbf{a}|T_i) = \prod_{j=1}^{p} P(a_j|T_i) \tag{7.7}$$

It is also often assumed that the prior probabilities of the taxa are equal, either because they are unknown or because they are variable. With this and the independence assumption, (7.5) for a vector of character values becomes

$$P(T_i|\mathbf{a}) = \frac{\displaystyle\prod_{j=1}^{p} P(a_j|T_i)}{\displaystyle\sum_{i=1}^{n} \prod_{j=1}^{p} P(a_j|T_i)} \tag{7.8}$$

Now the value of $P(T_i|\mathbf{a})$ could be determined for each of the n taxa and the unknown specimen assigned to the taxon for which this value is greatest. Here $P(T_i|\mathbf{a})$ is equivalent to what Willcox, Lapage & Holmes (1980) call an *identification score*. This method has been used to identify bacteria of medical importance. One merely chooses the taxon with the highest identification score. Despite the fact that the method does not take character correlations into account, it has been shown to be extremely powerful when about 30 characters are used for discrimination of taxa. If the identification score reaches a pre-assigned high level, for example 0.999, the identification is accepted as a successful one. If it is lower than this the identification is uncertain. An example from Willcox *et al.* (1980) is shown in Table 7.3.

If one is dealing with quantitative rather than qualitative characters, it is more usual to assume that they have a multivariate normal distribution rather than independence. This leads, for the special case of two taxa, to the well-known *linear discriminant function*, first suggested by Fisher (1936), and described in detail in section 7.3.1.

Returning for the present to a consideration of qualitative character states, one can think of a value for $P(T_i|\mathbf{a})$ (these are usually termed *posterior probabilities*) of, say, 0.95 as providing a boundary for the taxon T_i. One could, for example, calculate a similarity or distance to the centroid of a taxon (defined appropriately) that corresponds to this level of probability, and so define a *taxon radius*. Any specimen found within this radius is allocated to taxon T_i. Any

Table 7.3. *Computer identification based on Bayes' theorem (Willcox et al., 1980)–identified as* Pasteurella multocida

Results used in calculation (37 tests done):								
	(a)	(b)		(a)	(b)		(a)	(b)
Motility 37	−	1	MacConkey	−	25	Simmons Citr	−	1
Motility RT	−	1	Catalase	+	99	Urease	−	1
Growth 37	+	99	Oxidase	+	50	PPA	−	1
Growth RT	+	75	H and L Ferm	+	50	Glucose PWS	+	99
Pigment	−	1	Nitrate	+	99	Gas Glucose	−	1
Adonitol PWS	−	1	Lactose PWS	+	5	Sorbitol PWS	+	90
Arabinose PWS	−	15	Maltose PWS	−	1	Sucrose PWS	+	99
Cellobiose PWS	−	1	Mannitol PWS	+	90	Trehalose PWS	−	30
Dulcitol PWS	−	15	Raffinose PWS	−	5	Xylose PWS	−	50
Glycerol PWS	−	15	Rhamnose PWS	−	1	Starch PWS	−	1
Inositol PWS	−	1	Salicin PWS	−	1			
MR 37	−	1	VP 37	−	1	Indole	+	99
MR RT	−	1	VP RT	−	1			

Details of calculation:	
Group	Score
Pasteurella multocida	0.999 847
Pasteurella multocida (Atypical)	0.000 150

Column (a) tells one whether the reaction or test is regarded as positive or negative; column (b) the proportion (%) of tests that are for the species with the highest score. The extremes, 1% and 99%, refer to practically never, or practically always, positive, respectively.

specimen found not to lie within the taxon radius of any of the known taxa is regarded as unidentifiable. If an unknown falls just outside the taxon radius one would probably consider it to be an abnormal strain or variety of that taxon; if it were found to lie between two taxa it would be considered to be an intermediate form (perhaps a hybrid). One can use non-probabilistic concepts for taxon radius; but these will not be discussed here and the interested reader is referred to Sneath & Sokal (1973) and Sneath (1978b), where fuller discussions of *taxon radius models* may be found. Here we move on to consider Fisher's linear discriminant function.

7.3.1   *Fisher's linear discriminant function*

Let us suppose that now we have only two taxa, T_1 and T_2, and we wish to assign a new specimen to one of these on the basis of its scores p quantitative characters given in the vector $\mathbf{x}$. The assignment rule based on the posterior probabilities, $P(T_1|\mathbf{x})$ and $P(T_2|\mathbf{x})$, is as follows:

$$\text{if} \qquad \frac{P(T_1|\mathbf{x})}{P(T_2|\mathbf{x})} > 1 \qquad\qquad (7.9)$$

assign the specimen to T_1; otherwise assign it to T_2. If we now assume that the two taxa have equal prior probabilities, i.e. $P(T_1) = P(T_2)$, then from Bayes' theorem, as given in (7.5), we have

$$\frac{P(T_1|\mathbf{x})}{P(T_2|\mathbf{x})} = \frac{P(\mathbf{x}|T_1)}{P(\mathbf{x}|T_2)} \qquad\qquad (7.10)$$

If one now assumes that the distribution of $\mathbf{x}$ within taxa T_1 is multivariate normal with mean vector $\boldsymbol{\mu}_i$ and variance matrix, $\boldsymbol{\Sigma}$ (assumed to be the same in T_1 and T_2), then

$$\frac{P(\mathbf{x}|T_1)}{P(\mathbf{x}|T_2)} = \exp\left\{ -\tfrac{1}{2}[(\mathbf{x} - \boldsymbol{\mu}_1)'\boldsymbol{\Sigma}^{-1}(\mathbf{x} - \boldsymbol{\mu}_1) \right.$$
$$\left. - (\mathbf{x} - \boldsymbol{\mu}_2)'\boldsymbol{\Sigma}^{-1}(\mathbf{x} - \boldsymbol{\mu}_2)] \right\} \qquad (7.11)$$

Taking logarithms of the above leads to the assignment rule

$$\text{if} \qquad -\tfrac{1}{2}[(\mathbf{x} - \boldsymbol{\mu}_1)'\boldsymbol{\Sigma}^{-1}(\mathbf{x} - \boldsymbol{\mu}_1) - (\mathbf{x} - \boldsymbol{\mu}_2)'\boldsymbol{\Sigma}^{-1}(\mathbf{x} - \boldsymbol{\mu}_2)] > 0$$

assign the specimen to T_1; otherwise assign it to T_2. Rearrangement of this expression leads to

$$\mathbf{x}'\boldsymbol{\Sigma}^{-1}(\boldsymbol{\mu}_1 - \boldsymbol{\mu}_2) > \tfrac{1}{2}(\boldsymbol{\mu}_1 + \boldsymbol{\mu}_2)\boldsymbol{\Sigma}^{-1}(\boldsymbol{\mu}_1 - \boldsymbol{\mu}_2) \qquad (7.12)$$

and so the assignment rule becomes of the form

if $\mathbf{x}'\mathbf{w} > \theta$

assign specimen to T_1, otherwise assign it to T_2, where the weight vector, $\mathbf{w} = \Sigma^{-1}(\boldsymbol{\mu}_1 - \boldsymbol{\mu}_2)$, and the *threshold value*, $\theta = \frac{1}{2}(\boldsymbol{\mu}_1 + \boldsymbol{\mu}_2) \Sigma^{-1}(\boldsymbol{\mu}_1 - \boldsymbol{\mu}_2)$. This rule involves a *linear function* of the character values, which is generally known as the *discriminant function*.

So far it has been assumed that the mean vectors and the covariance matrix of the two taxa are known exactly. In practice, of course, this will not be so and they must be estimated from samples, one from each taxon; consequently, the weight vector and the threshold value are now estimated by

$$\hat{\mathbf{w}} = \mathbf{S}^{-1}(\bar{\mathbf{x}}_1 - \bar{\mathbf{x}}_2) \tag{7.13}$$

$$\hat{\theta} = \tfrac{1}{2}(\bar{\mathbf{x}}_1 + \mathbf{x}_2)\mathbf{S}^{-1}(\bar{\mathbf{x}}_1 - \bar{\mathbf{x}}_2) \tag{7.14}$$

where $\mathbf{S}$ is an estimate of the assumed common covariance matrix of the two taxa, and $\bar{\mathbf{x}}_1$ and $\bar{\mathbf{x}}_2$ are estimates of $\boldsymbol{\mu}_1$ and $\boldsymbol{\mu}_2$.

The weight vector $\hat{\mathbf{w}}$ was originally derived by Fisher in 1936 from a different starting point. He formulated the problem of discriminating between two classes in terms of choosing a linear function of the observations that has greatest variance between classes relative to variance within classes. Formulating the problem in this way enables the technique to be extended in a natural way to the situation where there are more than two taxa, when it is now known as canonical variate analysis.

7.3.2   *Canonical variate analysis*

Canonical variate analysis is similar in certain respects to principal components analysis. Again, transformed axes are sought but now the first axis is required to be in the direction of greatest variability between the means of the different taxa. The second axis is chosen to be orthogonal to the first and inclined in the direction of next greatest variability; similarly for the third and subsequent axes. The axes are termed canonical variates. In technical terms, the first canonical variate axis is that linear compound of the characters, $\mathbf{w}_1'\mathbf{x}$, which maximizes

$$\lambda_1 = \mathbf{w}_1'\mathbf{B}\mathbf{w}_1/\mathbf{w}_1'\mathbf{W}\mathbf{w}_1 \tag{7.15}$$

where $\mathbf{B}$ is the $(p \times p)$ between-taxa matrix of sums of squares and cross-products of the variables about their respective means, and $\mathbf{W}$ is the corresponding within-groups matrix. The second canonical variate is derived from the vector $\mathbf{w}_2'$, which maximizes

$$\lambda_2 = \mathbf{w}_2'\mathbf{B}\mathbf{w}_2/\mathbf{w}_2'\mathbf{W}\mathbf{w}_2 \tag{7.16}$$

subject to $\mathbf{w}_1'\mathbf{W}\mathbf{w}_2 = 0$, and so on.

It is easy to show that the coefficient vectors $\mathbf{w}_1', \mathbf{w}_2', \ldots$, etc., are given by the latent vectors of the matrix $\mathbf{B}\mathbf{W}^{-1}$. The number of vectors which can be extracted is given by the rank of the matrix and can be shown to be $\min\{p, n-1\}$, where p is the number of characters and n is the number of taxa. When $n = 2$ only a single canonical variate can be found and this is equivalent to Fisher's discriminant function. The latent roots of $\mathbf{B}\mathbf{W}^{-1}$ indicate how much between-taxa, relative to within-taxa, variability is associated with each canonical variate.

A set of canonical variate means for each taxon may now be found, and for a specimen requiring identification a set of canonical variate scores. The Euclidean distance of the specimen from each of the taxa, based on these scores, might now be determined and the specimen assigned to the taxon to which it is closest. (Such a procedure is essentially equivalent to computing the Mahalanobis' distance, $(\mathbf{x} - \bar{\mathbf{x}}_i)'\mathbf{W}^{-1}(\mathbf{x} - \bar{\mathbf{x}}_i)$, between the specimen and each taxon mean.)

A plot of taxon means in canonical variate space is often useful for displaying the relationships between taxa. Since the variates account for decreasing amounts of variance, a plot in the space of the first two or three variates is generally most useful.

7.3.3 *An example of canonical variate analysis*

The power of canonical variate analysis, especially in dealing with complex data obtained from chemical analysis of micro-organisms, will be illustrated by the work of MacFie, Gutteridge & Norris (1978) on aerobic food spoilage bacteria, and including that of O'Donnell, MacFie & Norris (1980) on three closely related species of *Bacillus*. These authors have generated the data used for identification by pyrolysis gas–liquid chromatography (p.g.l.c.). This is a process in which complex molecules are rapidly and reproducibly thermally degraded in an inert atmosphere (pyrolysis) and the products of this degradation separated and quantified (using gas–liquid chromatography). The results of p.g.l.c. can be plotted as pyrograms (see Figure 7.3) that can be considered as a characteristic 'finger-print' for a particular strain, provided that growth and chromatographic conditions are kept constant.

The groups of organisms used by MacFie *et al.* (1978) are shown in Table 7.4. These authors chose to measure the height of 24 of the chromatographic peaks (after standardization, to allow for changes in sample size; see Figure 7.3) and to treat these measurements as coordinates for the bacterial strains in a 24-dimensional Euclidean character space. Twenty-five replicate means were obtained for these measurements, to allow for variation between different p.g.l.c. runs. From these data, Mahalanobis' D^2 was calculated for every pair of the bacterial groups and the results are shown in Table 7.5.

Finally, canonical variate analysis was performed on these means,

Table 7.4. *Groups and numbers of strains of aerobic food spoilage bacteria used by MacFie* et al., *(1978)*

Group no.	Group name	No. of strains
1	*Moraxella*	5
2	*Pseudomonas*	7
3	*Lactobacillus*	5
4	*Microbacterium hermosphactum*	3
5	*Micrococcus*	5

Figure 7.3. Pyrogram of a strain of *Pseudomonas*. (Taken with permission from MacFie *et al.*, 1978.)

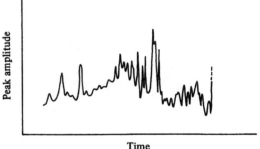

the first canonical variate plotted against the second for each group, and 95% confidence regions for the groups indicated (Figure 7.4). (A description of how these are obtained is given in Maxwell (1977), Chapter 9.) Clearly the analysis enables easy discrimination between these five groups of bacteria. These groups could easily have been differentiated by conventional bacteriological techniques, however,

Table 7.5. *Matrix of Mahalanobis* D^2 *among the bacterial groups given in Table* 7.4

Group no.	Group name	Group no.				
		1	2	3	4	5
1	*Moraxella*	0				
2	*Pseudomonas*	8	0			
3	*Lactobacillus*	12	15	0		
4	*Microbacterium hermosphactum*	50	50	58	0	
5	*Micrococcus*	18	24	17	53	0

Figure 7.4. Plot of the genus canonical variate means. Numbering as in Table 7.4. Circles indicate 95% confidence regions. (Taken with permission from MacFie *et al.*, 1978.)

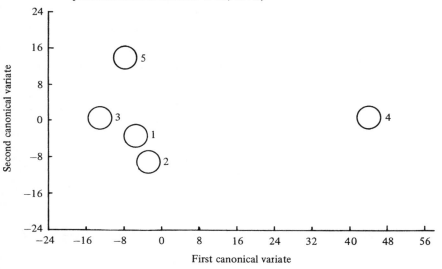

and it is of considerable interest to see how this method performs for more closely related organisms.

The second example is from the work of O'Donnell *et al.* (1980) on three closely related species of aerobic sporeformers, *Bacillus cereus* (9 strains), *Bacillus mycoides* (10 strains) and *Bacillus thuringiensis* (18 strains). *B. cereus* and *B. mycoides* are abundant soil bacteria, the latter producing distinctive colonies somewhat resembling those of a fungus. Apart from this property, these two species are so similar that *B. mycoides* has often been thought of as a variety of *B. cereus* (Stanier, Doudoroff & Adelberg, 1971). *B. thuringiensis* is an insect pathogen that produces a toxic protein crystal during sporulation. It is possible, however, to isolate mutants of *B. thuringiensis* that still form spores but have lost the ability to synthesize the toxic protein crystals. These mutants are no longer insect pathogens and cannot be distinguished from *B. cereus* by conventional methods. *B. thuringiensis* could also, therefore, be regarded as a variety of *B. cereus* (Stanier *et al.*, 1971). Another very closely related bacterium (*Bacillus anthracis*), not used by O'Donnell *et al.* (1980), is the causative organism for anthrax. Clearly it is of utmost importance to be able to identify these organisms quickly, efficiently and accurately.

Each strain was grown on nutrient agar until well sporulated and then subjected to p.g.l.c. Twenty-seven peak heights were measured from each resulting pyrogram and a canonical variate analysis carried out on the resulting data matrix (for this purpose replicated p.g.l.c.

Figure 7.5. Plot of species means relative to the first two canonical variate axes. (1) Mean of 18 strains of *B. thuringiensis*. (2) Mean of 9 strains of *B. mycoides*. (3) Mean of 10 strains of *B. cereus*. Circles define 95% confidence regions for each species. (Taken with permission from O' Donnell *et al.*, 1980.)

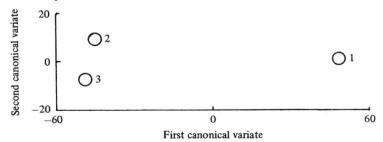

analyses representative of each strain were averaged). Figure 7.5 shows a plot of species means relative to the first two canonical variate axes. *B. cereus* and *B. mycoides* are close, but can easily be discriminated on the basis of the 27 peak heights. *B. thuringiensis* is well separated from the other two. O'Donnell *et al.* (1980) went on to investigate how few of these 27 measurements are needed for successful identification, and found that there were two measurements which could be used on their own to discriminate reasonably successfully between the three species (a single strain of *B. mycoides* being incorrectly assigned to *B. cereus*).

7.4   Summary

This chapter has been concerned with the identification of unknown OTUs; that is, the assignment of an unidentified specimen to a known taxon. There are basically two major approaches to this problem:

- (a) Sequential analysis of character states (for example in the construction and use of a diagnostic key), and
- (b) Simultaneous analysis of the character states (for example, in the use of a discriminant function or canonical variate analysis)

Use of methods belonging to the latter category is more likely to lead to the correct assignment and often requires measurements from fewer characters. These methods, however, frequently employ data obtained by the use of extensive or laborious laboratory experiments and so have been rarely used in practice, except by microbiologists. For further details of these methods, and for a summary of work done with other groups of organisms, the reader is referred to Sneath & Sokal (1973), Blackith & Reyment (1971) and Lachenbruch (1975). Readers are also referred to the excellent outline of discrimination techniques in Hand (1981), which presents an extensive survey of discriminant analysis methods as well as methods from the related field of pattern recognition.

8

The construction of evolutionary trees

8.1 Introduction

In this, the final chapter, ways in which phenetic information can be used for inferences concerning possible patterns of evolution will be introduced. Although these methods are not strictly part of numerical taxonomy, they are introduced to indicate to the student how one uses taxonomic data to generate scientific theories; that is, to show how numerical taxonomy is an integral part of systematics as a whole. As was stated in the first chapter, a numerical taxonomist makes inferences about evolution from phenetic data. He also bases his classifications on the phenetic properties of OTUs rather than on their supposed genealogies. This contrasts with the methods of taxonomists such as Mayr (1969) and, in particular, that of Hennig (1966).

It should also be clear, from what was said in Chapter 1, that the numerical taxonomist does not consider a classification to be a theory that can be tested by further taxonomic work. It is merely a way of summarizing data that may, or may not, be useful to future biologists. This does not mean that he cannot revise his classification after obtaining additional data. When one comes to making inferences about evolution from taxonomic data, however, one is trying to reconstruct actual historical sequences of events. Here one can quite easily be mistaken, and one can also try to find other evidence with which to test one's ideas. It is vital that the student understands this point. A dendrogram, for example, is merely a convenient, and perhaps useful, way of summarizing data, whereas an evolutionary tree is a summary of a scientific theory to be tested by further research.

8.2 Evolution as a branching process

The model of evolution used for constructing phylogenies is one of successive branchings, or successive splitting of populations into two or more subpopulations in which evolutionary changes then proceed independently. One need not assume that the rate of evolutionary change is constant throughout history, but many models are based on the assumption that, at any one time, the rate

Figure 8.1. The fifteen possible rooted trees for four taxa. (Taken with permission from Cavalli-Sforza & Bodmer, 1971.)

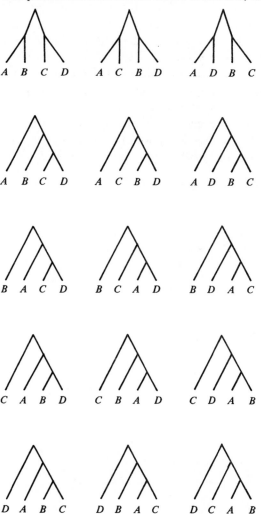

of evolution is the same in the different branches of the tree. In a more detailed model one can also specify the mechanisms responsible for these genetic changes; for example, mutation, migration, selection and drift. Various algorithms have been introduced for the construction of phylogenetic trees. These either produce trees with an apex (or root) or ones without (unrooted). The main difference between the two types of algorithm lies in the assumption of constant evolutionary rates in the different branches. Here examples of only the simplest methods will be discussed as an introduction to the subject. For further details the reader is referred to Sneath & Sokal (1973), Cavalli-Sforza & Edwards (1967) and Farris (1972).

Consider then four populations of OTUs, A, B, C' and D. There are fifteen possible ways in which these OTUs may have arisen from a single ancestral population by successive dichotomous splits. These are shown in Figure 8.1. If one ignores the root or apex the number of different trees is reduced to three, as is shown in Figure 8.2. The number of possible trees increases very rapidly with the

Figure 8.2. The three possible unrooted trees for four taxa.

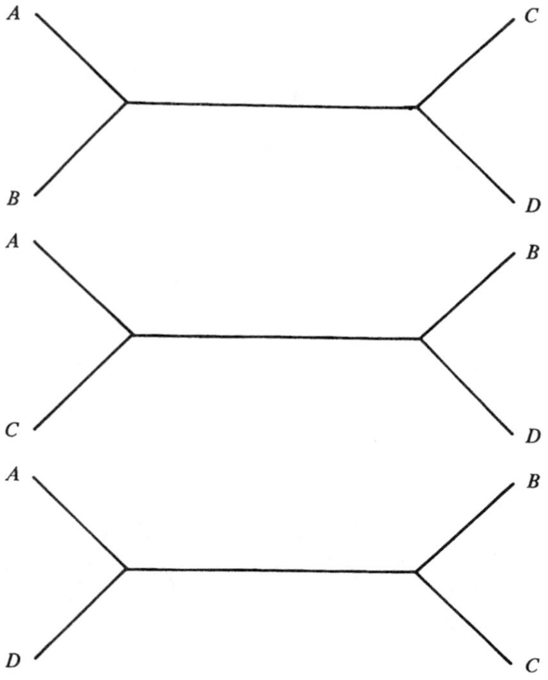

number of OTUs being studied. With three OTUs the number of rooted trees is three, and there is only a single unrooted one. With five OTUs these numbers are 90 and 15, respectively. Cavalli-Sforza & Edwards (1967) report that $(2t-3)!/[2^{t-2}(t-2)!]$ different rooted trees can be recognized for t OTUs. When $t = 10$, this equals 34,459, 425 trees. These authors also state that $(2t-5)!/[2^{t-3}(t-3)!]$ different unrooted trees are possible. This equals 2,027,025 when $t = 10$. The first problem in constructing a phylogenetic tree is the selection of the most plausible pathway from the very large number of possible ones. The next is the estimation of the positions of the branch points (nodes) and hence the lengths of the branches. The latter may, or may not, involve the reconstruction of the character- istics of the ancestors (*hypothetical taxonomic units*, HTUs) represent- ed by these nodes. Ideally one should be able to postulate a statistical model for the evolutionary processes that one is studying, and then compare the *likelihoods* for all possible trees, given the model and the observed characteristics of the OTUs. In practice, however, unless the model is very simple or there are very few OTUs to be studied, this methods of *maximum-likelihood* estimation is at present computa- tionally far too complex, even for high-speed computers (see Cavalli- Sforza & Edwards, 1967; and Edwards, 1970). The two-stage strategy given above can be thought of as a compromise that should produce a tree that is reasonably close to one that would have been generated by the full maximum-likelihood method.

8.3   The principle of minimal evolution

An idea that has been considered to be quite important by many evolutionists when attempting to construct phylogenetic trees is that of minimum evolution. It is not easy to justify its use, except that it is consistent with the intuitively satisfying principle of maxi- mum parsimony, but it produces trees that are quite close to those constructed using other criteria. Minimum evolution can be thought of as the minimum number of evolutionary steps (Camin & Sokal, 1965), the minimum number of mutational steps (Fitch & Margo- liash, 1967), or minimum tree length (Cavalli-Sforza & Edwards, 1967). When using this principle to construct trees one must decide whether characters should be ordered from evolutionarily primitive to advanced, and whether reversal of evolutionary processes is allowed. Here the discussion will be concerned with the construction of trees from data in which character states are not ordered and in which evolutionary reversals are allowed. Note that the idea of minimum evolution is also applicable to cases where only similarity or distance matrices are available, either because one is using published data which are in this form or because the character states cannot be recorded. An example of the latter case is where one is measuring biochemical similarities based on immunology or on nucleic acid hybridization.

Consider Figure 8.3(a). This shows one of the possible rooted trees from Figure 8.1 in which the three necessary HTUs have been added, as well as the lengths of all the branches. The total length is simply found by adding all of these distances together; that is $u + v + x + y + w + z$. For each of the different trees shown in Figure

Figure 8.3 (a) A rooted evolutionary tree for four OTUs. (b) An unrooted evolutionary tree for the same four OTUs.

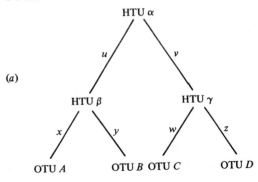

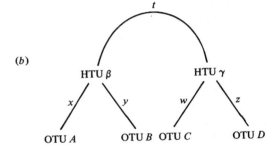

8.1 there will be an equivalent algebraic expression for its total length. The problem of finding the shortest evolutionary tree is then one of finding the shortest tree of each of the possible types (*topologies*) and then choosing the shortest of these. However, if there are more than a few OTUs being studied this is beyond the capacity of current computers. However, one can decide not to try to place HTU α and look for the minimum length unrooted tree (Figure 8.3(*b*)). It should be clear that this will have the same length as the minimum length rooted tree above, but will require much less computing power to find. But this will, again, be beyond the capabilities of current computers for all except trivial examples. In practice one uses an algorithm to produce a tree topology (perhaps with approximate estimates of the branch lengths) that is reasonably close to what one would expect from the complete solution, and then estimates the branch lengths accurately using a second algorithm. This procedure will yield a tree that is likely to be very close to the optimum solution, and may in fact be the best solution, but one can never be sure that a better tree does not exist.

8.4 The topology of the tree

How does one produce a genealogy that is a reasonable approximation to the true pathways of descent of the OTUs? If one assumes that evolution is basically divergent it might be quite sensible to use a dendrogram as a starting point. One assumes that OTUs that are similar, and are therefore grouped together by a clustering algorithm, have evolved from a relatively recent common ancestor. At this point, however, one has to think very carefully about the definition of a cluster that one has implicitly used in the construction of the clustering algorithm. The reader is referred to section 6.7. When using a dendrogram as the starting point for a phylogenetic tree, one merely uses the information from the topology of the graph and ignores the various fusion levels, and so on. One then uses one or more of the algorithms described in the next section to estimate the branch lengths of the tree. This produces an 'optimum' tree for a given topology or pattern of branching. One can, of course, try several different clustering algorithms and pick the resulting tree that is apparently the best of them all.

There are also other algorithms, differing from those described in

Chapter 6, that can be used to generate a reasonable topology. Some of these are discussed in detail by Farris (1970, 1972), and the interested reader is referred to these papers for a fuller discussion of the problems involved. Here one method only will be described. This produces a *minimum spanning tree* (Prim, 1957; Edwards & Cavalli-Sforza, 1964; see section 5.4) which may be computed as follows (Farris, 1970):

1. Pick an OTU, say Q, as a starting point. It does not matter which OTU is used. Go to 2.
2. Find the OTU that is closest to Q. Join it to Q to form a graph with a single connection. Go to 3.
3. Compute the difference between each unplaced OTU and the graph. The difference between an OTU, A, and a graph is defined as the minimum distance between A and the OTUs connected by the graph. Go to 4.
4. Find the OTU that is closest to the graph. Add it to the graph by connecting it to the OTU from which it differs least. Go to 5.
5. If any OTUs remain unplaced, go to 3. Otherwise, stop.

8.5   Optimization of trees

Having produced an approximation of the required evolutionary tree, either by using a clustering algorithm or some other tree-producing method, one can then proceed to estimate its branch lengths using one of a few criteria to assess goodness-of-fit to the taxonomic data. If one were to produce a sensible statistical model for the evolutionary process one could then use maximum likelihood estimation (Cavalli-Sforza & Edwards, 1967). One could also use minimum tree length as an optimality criterion. Algorithms for minimum length trees given a particular tree topology are described by Fitch (1971), Farris (1970, 1972), and Hartigan (1973). These will not be explained here; instead, a different criterion of optimality will be introduced.

If one has a reasonable measure of the taxonomic distance between two independently evolving populations, one can assume that the measure will increase with time (evolution being essentially divergent). In addition, the distances produced in a given time will simply add on to those produced in a previous time interval. Consider Figure 8.3(*b*). The *patristic* distance between two OTUs is defined as that distance between them which is measured by the appropriate branch lengths of the evolutionary tree (Farris, 1972). So, for example, the patristic distance between OTU *A* and OTU *B* is $x + y$, and that between OTU *A* and OTU *D* is $x + t + z$. The *phenetic* distance is that provided by the taxonomic data matrix. One can construct a phylogenetic tree in which the patristic distances are as close as possible to the observed phenetic distances. From Figure 8.3(*b*) one has:

$$d_{AB} = x + y + \text{`error'}$$

$$d_{AC} = x + t + w + \text{`error'}$$

$$d_{AD} = x + t + z + \text{`error'}$$

$$d_{BC} = y + t + w + \text{`error'}$$

$$d_{BD} = y + t + z + \text{`error'}$$

$$d_{CD} = w + z + \text{`error'}$$

where, for example, d_{AB} is the phenetic distance between OTUs *A* and *B*. The 'error' terms represent the differences between the observed phenetic distances and the computed patristic ones. One can estimate the terms x, y, z, and so on, by minimizing the sum of the squared errors—the method of 'least squares' (Cavalli-Sforza & Edwards, 1967). Alternatively, one might wish to minimize the sum of the absolute values of the errors (Fitch & Margoliash, 1967). Either statistic can be used as a measure of the goodness-of-fit of the resulting tree to the taxonomic data.

8.6   Reticulate evolution: the problem of hybrids

The methods that have been described in this chapter are based on the fundamental assumption that lineages may branch but never fuse. Fusion of lineages is called reticulate evolution, and it introduces difficulties for the use of phenetic data in making inferences about evolutionary trees. Consider Figure 8.4(*a*) which shows a hypothetical dendrogram for four OTUs, *A* to *D*. On the assumption that there has been no hybridization during the evolution of these OTUs, one can produce an evolutionary tree with the topology of Figure 8.4(*b*). Suppose, however, that one has evidence, say, from the cytology of the OTUs, that OTU *B* has evolved from a hybrid formed from the recent ancestors of OTU *A* and OTU *C*. This would imply that the evolutionary tree in Figure 8.4(*b*) is of little use, despite the fact that a classification based on the dendrogram in Figure 8.4(*a*) might be quite satisfactory. Half of the genome of OTU *B* will have evolved along a pathway similar to that in Figure 8.4(*b*) and the rest by a pathway indicated by Figure 8.4(*c*). A realistic evolutionary tree for the four OTUs might have a form similar to that shown in Figure 8.4(*d*). (See Sneath (1975) for similar representations.)

How common is this problem in work on evolutionary biology? Reticulate evolution is thought to be rare in animals, but in plants hybridization frequently occurs between different taxa, leading to persisting and evolving lineages (Davis & Heywood, 1963; Grant, 1971). Hybridization usually occurs via the formation of allopoly-

Figure 8.4. (*a*) Dendrogram for four hypothetical OTUs.
(*b*) One possible evolutionary tree for these four OTUs.
(*c*) An alternative evolutionary tree for the four OTUs.
(*d*) An evolutionary tree to illustrate the occurrence of hybridization.

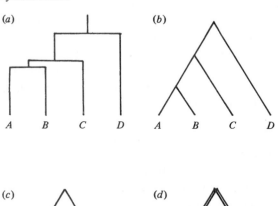

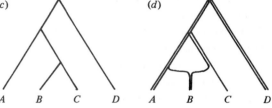

ploids in sexually reproducing plants, and in some genera there is a large proportion of species that are thought to have arisen in this way (Davis & Heywood, 1963). The genus *Solanum* is a well-known example, and the reader is referred to the work of Edmonds (1977, 1978) on the numerical taxonomy of OTUs from this genus. There is also evidence that reticulate evolution is common in micro-organisms (see Jones & Sneath, 1970, but also section 8.7). Finally, there is increasing evidence that the eukaryotic cell itself has evolved from an ancient hybrid derived from several primitive cell lines (see Schwartz & Dayhoff, 1978, and, again, section 8.7).

8.7 Gene phylogenies

This section will be mainly concerned with inferences that can be made from the study of amino acid sequences in proteins, or of nucleotide sequences in nucleic acids. In the second chapter an operational definition of homology was given, and two nucleotide sequences were compared as an example of how this definition could be used. Now an additional complication will be introduced to show that the simple operational definition of homology is inadequate in many situations. When considering sequence data it is essential to distinguish between two kinds of homology. Following a gene duplication, the two resulting genes may evolve independently, perhaps developing different functions within the cell, while descending side by side in the same phyletic lineage. These genes are said to be *paralogous*. As an example, consider the genes for myoglobin and alpha haemoglobin. They are thought to have evolved from a common ancestral gene that duplicated very early in the evolution of vertebrates, and they have subsequently evolved independently within each evolutionary line of animals. Although myoglobin from a chicken and alpha haemoglobin from man are homologous, it would be absurd to compare the two sequences in order to gain information on the relationship between the two taxa. One should either compare sequences of myoglobin from the two species, or haemoglobin sequences. One must compare directly only those sequences whose genes have a lineage that precisely corresponds, in a one-to-one fashion, to the descent of the species in which they are found. These genes are called *orthologous*. There is no simple operational definition, similar to that given for homology in Chapter

2, that will distinguish paralogous from orthologous sequences.

So, one can look at patterns of descent by comparison of orthologous sequences or of paralogous sequences of macromolecules (the latter usually taken from a single taxon). In either case one can use numerical methods to produce evolutionary trees for the genes involved, but only the gene phylogenies for the orthologous

Figure 8.5. Globin gene phylogeny. Numbers on segments are nucleotide replacements required to account for the descent of the five present-day sequences from a common ancestor. (Taken with permission from Fitch and Margoliash, 1970.)

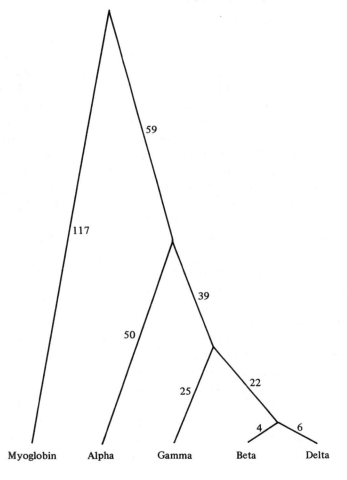

case will yield any information about the descent of the taxa in which they are found. As an example of the results of analyses of paralogous globin sequences, consider Figure 8.5. This shows a history of gene duplications. An ancestral globin gene duplicated; following this, the product of one of the resulting genes developed into a muscle oxygen carrier (myoglobin) while the product of the other became the blood oxygen carrier (haemoglobin). Further gene duplications in the haemoglobin line led to the evolution of alpha and beta haemoglobins, so that the tetrameric $\alpha_2\beta_2$-form of haemoglobin became possible. The gene for the beta chain then duplicated again to permit the development of the gamma chain found in foetal haemoglobins, and so on.

If one were to produce a phylogenetic tree for taxa (or a dendrogram, if one were merely interested in classification) using sequence data from haemoglobins, would one expect to obtain a tree very similar to one based on myoglobin sequences? Since reticulate evolution is not thought to have occurred in vertebrates, the answer ought to be 'yes'. Similarly, one would expect these trees to be fairly close to phylogenies produced by traditional methods. If one were to study plant or microbial macromolecules, however, this might not be the case. Bacterial genes can be transferred between quite distantly related taxonomic groups via phages or plasmids and this type of transfer may have been quite common during the evolution of present-day micro-organisms. The complications that this introduces into bacterial taxonomy have been illustrated by Jones & Sneath (1970):

> If these views are accepted, it can be seen that a strain X of taxospecies A, carrying a plasmid P derived from the genome of taxospecies B, has multiple relationships to species B. These include genetic relationship (in the wide sense) due to plasmid transfer. But X and B also have a genomic relationship (by virtue of the genes on the plasmid), a phenetic relationship (due to the genes on P), and in addition a cladistic relationship due to the origin of P. Other strains of species A have a close relationship to strain X, a relationship which is at least phenetic (but which may also be assumed to be also genomic and cladistic). However, they have only a distant relationship of any kind to species B, except through the intermediacy of strain X.

Recent data on cytochrome *c* sequences from the Rhodospirillaceae or purple non-sulphur photosynthetic bacteria has suggested a classification, and subsequent evolutionary tree, that is not consistent with classifications produced from traditional taxonomic evidence (Ambler *et al.,* 1979*a,b*). It has been suggested that macromolecular information might be of little use in deciphering bacterial phylogeny because of widespread transfer of genes and subsequent 'scrambling' of the genetic information. One way of testing whether this suggestion might be correct for the purple photosynthetic bacteria would be to produce a second classification, based on sequence data from another macromolecule. Woese, Gibson & Fox (1980) have done this using

Figure 8.6. The relationships among various purple photosynthetic bacteria as determined by (*a*) 16 S ribosomal RNA sequence comparisons and (*b*) cytochrome *c* sequence comparisons. (Taken with permission from Woese, Gibson and Fox, 1980).

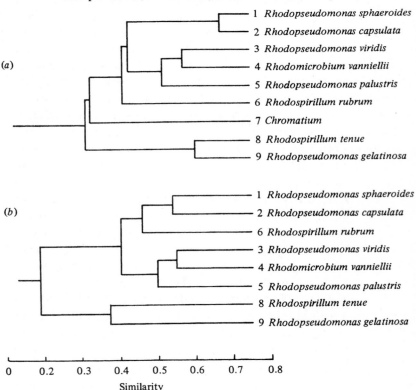

the sequences of 16 S ribosomal RNA. Their results are shown in Figure 8.6. The two dendrograms are remarkably similar, suggesting that gene transfer should not be held responsible for the conflict between the classifications based on sequence data and those obtained by traditional means.

An approach which one might wish to use would be to assume that reticulate evolution has indeed occurred, and to use sequence data as a probe to find out how and when it might have happened. Chloroplasts of green plants contain, like prokaryotic cells, 70 S ribosomes that differ from the 80 S ribosomes of the eukaryotic cytoplasm. This, along with other evidence, has led to the hypothesis that modern chloroplasts are descendants of endosymbiotic prokaryotes, similar to cyanobacteria, which, early in evolution, entered primitive eukaryotic cells (see Schwartz & Dayhoff, 1978). This, and a similar hypothesis concerning the origin of eukaryotic mitochondria, can be tested by comparison of sequences of macromolecules from prokaryotes, eukaryotic organelles and eukaryotic cytoplasms. Schwartz & Dayhoff (1978) summarize the analysis of sequences of ferredoxins, *c* cytochromes and 5 S RNAs, which, they claim, support the theory that eukaryotic organelles are descended from prokaryotic ancestors. Many of their conclusions, however, have been challenged (Demoulin, 1979). The sequences of human mitochondrial ribosomal RNA genes are significantly homologous in some regions to both eukaryotic (i.e. cytoplasmic) and prokaryotic sequences, but are distinctive enough to suggest that mitochondria are not descended from recognizable relatives of present-day organisms (Eperon, Anderson & Nierlich, 1980). On the other hand, 16 S rRNA from maize chloroplasts is very similar to that from the bacterium *Escherichia coli*, supporting the idea that chloroplasts and prokaryotic cells do, indeed, have a common origin (Schwarz & Kössel, 1980). This conclusion is supported by the work of Fox *et al.* (1980), who have provided a dendrogram to illustrate the similarities between 16 S rRNAs from chloroplasts and cyanobacteria (Figure 8.7).

What else can be learnt from sequence data? Provided that one can estimate evolutionary rates from fossil evidence, one can calibrate the speeds of macromolecular evolution and compare them with, say, evolutionary changes in morphology (Fitch, 1976; Wilson, Carlson & White, 1977). Are evolutionary rates constant? Are evolutionary rates of change for one type of macromolecule the same as for another? Given their functional importance, do particular sequences within a macromolecule change at much slower rates than others? These and many other questions will be answered with the development of macromolecular taxonomy (relying, almost entirely, on numerical methods of analysis) and its subsequent incorporation into systematics as a whole.

8.8   **Summary**

This chapter contains a brief outline of the way in which numerical methods can be used in the construction of phylogenetic trees. In addition, it is hoped that the discussion illustrates the way in which the results of numerical taxonomic studies can be used in a

Figure 8.7. Topology of the provisional phylogenetic tree for cyanobacteria and chloroplasts. (Taken with permission from Fox *et al.*, 1980.)

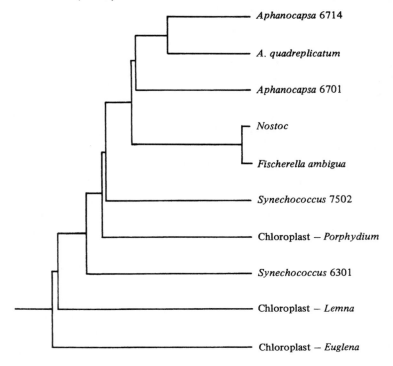

Aphanocapsa 6714

A. quadreplicatum

Aphanocapsa 6701

Nostoc

Fischerella ambigua

Synechococcus 7502

Chloroplast — *Porphydium*

Synechococcus 6301

Chloroplast — *Lemna*

Chloroplast — *Euglena*

more general study of diversity (i.e. systematics). Unlike classifications, inferences concerning patterns of evolution can be wrong, and the results given in this chapter as examples must be regarded as tentative. The construction of 'better' evolutionary trees will, presumably, have to await the development of more realistic models of evolutionary processes, but one would expect that the mathematical methods used might be similar to the maximum likelihood procedures of Cavalli-Sforza & Edwards (1967). A more comprehensive discussion of this problem is given by Thompson (1975).

REFERENCES

Ambler, R. P., Daniel, M., Hermoso, J., Meyer, T. E., Bartsch, R. E. & Kamen, M. D. (1979a) Cytochrome c_2 sequence variation among the recognised species of purple non-sulphur photosynthetic bacteria. *Nature*, **278**, 659–60.

Ambler, R. P., Meyer, T. E. & Kamen, M. D. (1979b) Anomalies in amino acid sequences of small cytochromes c and cytochromes c' from two species of purple photosynthetic bacteria. *Nature*, **278**, 661–2.

Anderberg, M. R. (1973) *Cluster Analysis for Applications*. New York: Academic.

Andrews, D. F. (1972) Plots of high dimensional data. *Biometrics*, **28**, 125–36.

Austin, M. T. (1972) Models and analysis of descriptive vegetation data. In *Mathematical Models in Ecology*, ed. J. N. R. Jeffers, pp. 61–86. Oxford: Blackwell.

Baum, B. R. (1977) Reduction of dimensionality for heuristic purposes. *Taxon*, **26**, 191–5.

Bennett, S. & Bowers, D. (1976) *An Introduction to Multivariate Techniques for Social and Behavioural Sciences*. London: Macmillan.

Bessey, C. E. (1908) The taxonomic aspect of the species. *American Naturalist*, **42**, 218–24.

Blackith, R. E. & Reyment, R. A. (1971) *Multivariate Morphometrics*. London: Academic.

Blashfield, R. K. (1976) Mixture model tests of cluster analysis: accuracy of four hierarchical agglomerative methods. *Psychological Bulletin*, **83**, 377–88.

Boyce, A. J. (1969) Mapping diversity. A comparative study of some numerical methods. In *Numerical Taxonomy*, ed. A. J. Cole, pp. 1–30. New York: Academic.

Brown, P. J. (1977) Functions for selecting tests in diagnostic key construction. *Biometrika*, **64**, 589–96.

Cain, A. J. & Harrison, G. A. (1958) An analysis of the taxonomist's judgement of affinity. *Proceedings of the Zoological Society of London*, **131**, 85–98.

Camin, J. H. & Sokal, R. R. (1965) A method for deducing branching sequences in phylogeny. *Evolution*, **19**, 311–326.

Carmichael, J. W. & Sneath, P. H. A. (1969) Taxometric Maps. *Systematic Zoology*, **18**, 402–15.

Cavalli-Sforza, L. L. & Bodmer, W. F. (1971) *The Genetics of Human Populations*. San Francisco: Freeman.

Cavalli-Sforza, L. L. & Edwards, A. W. F. (1967) Phylogenetic analysis–models and estimation procedures. *American Journal of Human Genetics*, **19**, 233–57.

Cheetham, A. H. & Hazel, J. E. (1969) Binary (presence–absence) similarity coefficients. *Journal of Paleontology*, **43**, 1130–36.

Clifford, D. H. T. & Stephenson, W. (1975) *An Introduction to Numerical Classification.* New York: Academic.

Cole, A. J. & Wishart, D. (1970) An improved algorithm for the Jardine–Sibson method of generating overlapping clusters. *Computer Journal,* 13, 156–63.

Cormack, R. M. (1971) A review of classification. *Journal of the Royal Statistical Society, Series A,* 134, 321–67.

Cunningham, K. M. & Ogilvie, J. C. (1972) Evaluation of hierarchical grouping techniques. A preliminary study. *Computer Journal,* 15, 209–13.

Czekanowski, J. (1909) Zur Differentialdiagnose der Neandertalgruppe. *Korrespondenzblatt der Deutschen Geselschaft für Anthropologie, Ethnologie und Urgeschichte,* 40, 44–7.

Czekanowski, J. (1932) 'Coefficient of racial likeness' and 'durchschnittliche Differenz.' *Anthropologischer Anzeiger,* 9, 227–49.

Davis, P. H. & Heywood, V. H. (1963) *Principles of Angiosperm Taxonomy.* Edinburgh: Oliver & Boyd.

Demoulin, V. (1979) Protein and nucleic acid sequence data and phylogeny. *Science,* 205, 1036–38.

Dubes, R. & Jain, A. K. (1979) Validity studies in clustering methodologies. *Pattern Recognition Journal,* 11, 235–54.

Eades, D. C. (1965) The inappropriateness of the correlation coefficient as a measure of taxonomic resemblance. *Systematic Zoology,* 14, 98–100.

Edmonds, J. M. (1977) Taxonomic studies on *Solanum* L. section *Solanum* (*Maurella*). *Botanical Journal of the Linnean Society of London,* 75, 141–78.

Edmonds, J. M. (1978) Numerical taxonomic studies on *Solanum* L. section *Solanum* (*Maurella*). *Botanical Journal of the Linnean Society of London,* 76, 27–51.

Edwards, A. W. F. (1970) Estimating the branch points of a diffusion process. *Journal of the Royal Statistical Society, Series B,* 32, 155–74.

Edwards, A. W. F. & Cavalli-Sforza, L. L. (1964) Reconstruction of evolutionary trees. In *Phenetic and Phylogenetic Classification,* ed. V. H. Heywood and J. McNeill, pp. 67–76. Systematics Association Publication 6. London: Systematics Association.

Englemann, L. & Hartigan, J. A. (1969) Percentage points of a test for clusters. *Journal of the American Statistical Association.* 64, 1647–48.

Eperon, I. C., Anderson, S. & Nierlich, D. P. (1980) Distinctive sequence of human mitochondrial ribosomal RNA genes. *Nature,* 286, 460–67.

Everitt, B. S. (1978) *Graphical Techniques for Multivariate Data.* London: Heinemann.

Everitt, B. S. (1980) *Cluster Analysis* (2nd edn). London: Heinemann.

Everitt, B. S. & Hand, D. J. (1981) *Finite Mixture Distributions.* London: Chapman & Hall.

Farris, J. S. (1970) Methods for computing Wagner trees. *Systematic Zoology,* 19, 83–92.

Farris, J. W. (1972) Estimating phylogenetic trees from distance matrices. *American Naturalist,* 106, 645–68.

Feyerabend, P. (1975) *Against Method.* London: Verso.

Fisher, D. R. (1968) A study of faunal resemblance using numerical taxonomy and factor analysis. *Systematic Zoology,* 17, 48–63.

Fisher, R. A. (1936) The use of multiple measurements in taxonomic problems. *Annals of Eugenics,* **7,** 179–88.

Fisher, W. D. (1969) *Clustering and Aggregation in Economics.* Baltimore: Johns Hopkins University Press.

Fitch, W. M. (1970) Further improvements in the method of testing evolutionary homology among proteins. *Journal of Molecular Biology,* **49,** 1–14.

Fitch, W. M. (1971) Towards defining the course of evolution. Minimum change for a specific tree topology. *Systematic Zoology,* **20,** 406–16.

Fitch, W. M. (1976) Molecular evolutionary clocks. In *Molecular Evolution,* ed. F. J. Ayala, pp. 160–78. Sunderland, Mass.: Sinauer.

Fitch, W. M. & Margoliash, E. (1967) Construction of phylogenetic trees. *Science,* **155,** 279–84.

Fitch, W. M. & Margoliash, E. (1970) The usefulness of amino acid and nucleotide sequences in evolutionary studies. *Evolutionary Biology,* **4,** 67–109.

Fleiss, J. L. & Zubin, J. (1969) On the methods and theory of clustering. *Multivariate Behavioural Research,* **4,** 235–50.

Fox, G. E., Stackebrandt, E., Hespell, R. B., Gibson, J., Maniloff, J., Dyer, T. A., Wolfe, R. S., Balch, W. E., Tanner, R. S., Magrum, L. J., Zablen, L. B., Blakemore, R., Gupta, R., Bonen, L., Lewis, B. J., Stahl, D. A., Luehrsen, K. R., Chen, K. N. & Woese, C. R. (1980) The phylogeny of prokaryotes. *Science,* **209,** 457–63.

Garey, M. R. (1972) Optimal binary identification procedures. *SIAM Journal of Applied Mathematics,* **23,** 173–86.

Garey, M. R. & Graham, R. L. (1974) Performance bounds on the splitting algorithm for binary testing. *Acta Informatica,* **3,** 347–55.

Gilmour, J. S. L. (1940) Taxonomy and philosophy. In *The New Systematics,* ed. J. S. Huxley, pp. 461–72. Oxford: Clarendon.

Gnanadesikan, R., Kettenring, J. R. & Landwehr, J. M. (1977) Interpreting and assessing the results of cluster analysis. *Bulletin of the International Statistical Institute,* **XLV11(2),** 451–63.

Gower, J. C. (1966) Some distance properties of latent root and vector methods used in multivariate analysis. *Biometrika,* **53,** 325–38.

Gower, J. C. (1971) A general coefficient of similarity and some of its properties. *Biometrics,* **27,** 857–72.

Gower, J. C. (1975) Goodness-of-fit criteria for classification and other patterned structures. In *Proceedings of the 8th International Conference on Numerical Taxonomy,* ed. G. Estabook, pp. 38–62. London: Freeman.

Gower, J. C. & Payne, R. W. (1975) A comparison of different criteria for selecting binary tests in diagnostic keys. *Biometrika,* **62,** 665–72.

Gower, J. C. & Ross, G. J. S. (1969) Minimum spanning trees and single linkage cluster analysis. *Applied Statistics,* **18,** 54–64.

Grant, V. E. (1971) *Plant Speciation.* New York: Columbia University Press.

Haltenorth, T. (1937) Die verwandtschaftliche Stellung der Grosskatzen zu einander. *Zeitschift für Säugetierkunde,* **12,** 97–240.

Hand, D. J. (1981) *Discrimination and Classification.* London: Wiley.

Hartigan, J. A. (1973) Minimum mutation fits to a given tree. *Biometrics,* **29,** 53–65.

Hartigan, J. A. (1975) *Clustering Algorithms.* New York: Wiley.

Hays, W. L. (1973) *Statistics for the Social Sciences*. New York: Holt, Reinhart & Winston.

Hennig, W. (1966) *Phylogenetic Systematics*. Urbana: University of Illinois Press.

Jaccard, P. (1908) Nouvelles recherches sur la distribution florale. *Bulletin de la Société Vaudoise de Science Naturelle*, **44**, 223–70.

Jardine, N. (1967) The concept of homology in biology. *British Journal of the Philosophical Society*, **18**, 125–39.

Jardine, N. & Sibson, R. (1968) The construction of hierarchic and non-hierarchic classifications. *Computer Journal*, **11**, 117–84.

Jardine, N. & Sibson, R. (1971) *Mathematical Taxonomy*. London: Wiley.

Jeffers, J. N. R. (1967) Two case studies in the application of principal component analysis. *Applied Statistics*, **16**, 225–36.

Johnson, S. C. (1967) Hierarchical clustering schemes. *Psychometrika*, **32**, 241–54.

Jolicoeur, P. & Mosimann, J. E. (1960) Size and shape variation in the painted turtle: A principal components analysis. *Growth*, **24**, 339–54.

Jones, D. & Sneath, P. H. A. (1970) Genetic transfer and bacterial taxonomy. *Bacteriological Reviews*, **34**, 40–81.

Kendall, D. G. (1971) Seriation from abundance matrices. In *Mathematics in the Archaeological and Historical Sciences*, ed. F. R. Hodson, D. G. Kendall and P. Tautu, pp. 215–52. Edinburgh: Edinburgh University Press.

Kendall, D. G. (1975) Review lecture: The recovery of structure from fragmentary information. *Philosophical Transactions of the Royal Society, Series A*, **279**, 547–82.

Kruskal, J. B. (1964a) Multidimensional scaling by optimizing goodness of fit to a non-metric hypothesis. *Psychometrika*, **29**, 1–27.

Kruskal, J. B. (1964b) Non-metric multidimensional scaling: a numerical method. *Psychometrika*, **29**, 115–29.

Kruskal, J. B. & Carroll, J. D. (1969) Geometric models and badness-of-fit functions. In *Multivariate Analysis*, vol. 2, ed. P. R. Krishnaiah, pp. 639–71. New York: Academic.

Kruskal, J. B. & Wish, M. (1978) *Multidimensional Scaling*. California: Sage.

Kuiper, F. K. & Fisher, L. (1975) A Monte Carlo comparison of six clustering procedures. *Biometrics*, **31**, 777–83.

Lachenbruch, P. A. (1975) *Discriminant Analysis*. New York: Hafner.

Lambert, J. M. & Williams, W. T. (1962) Multivariate methods in plant ecology, IV. Nodal Analysis. *Journal of Ecology*, **50**, 775–802.

Lambert, J. M. & Williams, W. T. (1966) Multivariate methods in plant ecology, V. Comparison of information analysis and association analysis. *Journal of Ecology*, **54**, 635–64.

Lance, G. N. & Williams, W. T. (1967) A general theory of classificatory sorting strategies: 1. Hierarchical systems. *Computer Journal*, **9**, 373–80.

Ling, R. F. (1972) On the theory and construction of k-clusters. *Computer Journal*, **15**, 326–32.

Ling, R. F. & Killough, G. S. (1976) Probability tables for cluster analysis based on a theory of random graphs. *Journal of the American Statistical Association*, **71**, 293–300.

Lockhart, W. R. (1970) Coding the data. In *Methods for Numerical Taxonomy*, ed. W. R. Lockhart and J. Lislon. Bethesda, Md: American Society of Microbiology.

MacFie, H. J. H., Gutteridge, C. S. & Norris, J. R. (1978) Use of canonical variates analysis in differentiation of bacteria by pyrolysis gas–liquid chromatography. *Journal of General Microbiology*, **104**, 67–74.

MacNaughton-Smith, P. (1965) *Some Statistical and other Numerical Techniques for Classifying Individuals*. Home Office Research Unit Report No. 6. London: H.M.S.O.

Mardia, K. V., Kent, J. T. & Bibby, J. M. (1979) *Multivariate Analysis*. London: Academic.

Maxwell, A. W. E. (1977) *Multivariate Analysis in Behavioural Research*. London: Chapman & Hall.

Mayr, E. (1969) *Principles of Systematic Zoology*. New York: McGraw-Hill.

Michener, C. D. (1970) Diverse approaches to systematics. *Evolutionary Biology*, **4**, 1–38.

Minkoff, E. C. (1965) The effects on classification of slight alterations in numerical technique. *Systematic Zoology*, **14**, 196–213.

Mojena, R. (1977) Hierarchical grouping methods and stopping rules: an evaluation. *Computer Journal*, **20**, 359–63.

Morrison, D. F. (1967) *Multivariate Statistical Methods*. New York: McGraw-Hill.

Morrison, D. G. (1967) Measurement problems in cluster analysis. *Management Science*, **13**, 775–80.

Morse, L. E. (1971) Specimen identification and key construction with time sharing computers. *Taxon*, **20**, 269–82.

Moss, W. W. (1967) Some new analytic and graphic approaches to numerical taxonomy, with an example from the Dermanyssidae (Acari) *Systematic Zoology*, **16**, 177–207.

O'Donnell, A. G., MacFie, H. J. H. & Norris, J. R. (1980) An investigation of the relationship between *Bacillus cereus*, *Bacillus thuringiensis* and *Bacillus mycoides* using Pyrolysis Gas–Liquid Chromatography. *Journal of General Microbiology*, **119**, 189–94.

Ore, O. (1963) *Graphs and their Uses*. New York: Random House.

Pankhurst, R. J. (1970) A computer program for generating diagnostic keys. *New Phytologist*, **62**, 35–43.

Pankhurst, R. J. (1978) *Biological Identification. The Principles and Practice of Identification Methods in Biology*. London: Edward Arnold.

Payne, R. W. (1975) Genkey: a program for constructing diagnostic keys. In *Biological Identification with Computers*. Systematics Association Special Volume. No. 7, ed. R. J. Pankhurst, pp. 65–72. London: Academic.

Payne, R. W. & Preece, D. A. (1977) Incorporating checks against observer error into identification keys. *New Phytologist*, **79**, 203–9.

Payne, R. W. & Preece, D. A. (1980) Identification keys and diagnostic tables – a review. *Journal of the Royal Statistical Society, Series A*, **143**, 253–82.

Penrose, L. S. (1954) Distance, size and shape. *Annals of Eugenics*, **28**, 337–43.

Picard, C. F. (1965) *Théorie des Questionnaires*, Paris: Gauthier-Villars.

Picard, C. F. (1980) *Graphs and Questionnaires*. Mathematical Studies, Volume 32. Amsterdam: North Holland.

Prentice, H. C. (1979) Numerical analysis of infraspecific variation in European *Silene alba* and *S. dioica* (Caryophyllaceae). *Botanical Journal of the Linnaean Society*, **78**, 181–212.

Prentice, H. C. (1980) Variation in *Silene dioica* (L) Clairv. Numerical analysis of populations from Scotland. *Watsonia*, **13**, 11–26.

Prim, R. C. (1957) Shortest connection matrix network and some generalisations. *Bell System Technical Journal*, **36**, 1389–401.

Rao, C. R. (1964) The use and interpretation of principal components analysis in applied research. *Sankhyā*, **26**, 329–58.

Rohlf, F. J. (1967) Correlated characters in numerical taxonomy. *Systematic Zoology*, **16**, 109–26.

Rohlf, F. J. (1968) Stereograms in numerical taxonomy. *Systematic Zoology*, **17**, 246–55.

Rohlf, F. J. (1970) Adaptive hierarchical clustering schemes. *Systematic Zoology*, **19**, 58–82.

Rohlf, F. J. & Fisher, D. L. (1968) Test for hierarchical structure in random data sets. *Systematic Zoology*, **17**, 407–12.

Rohlf, F. J. & Sokal, R. R. (1965) Coefficients of correlation and distance in numerical taxonomy. *University of Kansas Scientific Bulletin*, **45**, 3–27.

Ruse, M. (1973) *The Philosophy of Biology*. London: Hutchinson.

Sammon, J. W. (1969) A non-linear mapping for data structure analysis. *IEEE Transactions on Computers*, **C18**, 401–9.

Sandvik, L. (1976) A note on the theory of dichotomous keys. *New Phytologist*, **76**, 555–8.

Saunders, R. & Funk, G. M. (1977) Poisson limits for a clustering model of Strauss. *Journal of Applied Probability*, **14**, 776–84.

Schwartz, R. M. & Dayhoff, M. O. (1978) Origins of prokaryotes, mitochondria and chloroplasts. *Science*, **199**, 395–403.

Schwarz, Z. & Kössel, H. (1980) The primary structure of 16 S rRNA from *Zea mays* chloroplast is homologous to *E. coli* 16 S rRNA. *Nature*, **283**, 739–42.

Shannon, C. E. (1948) A mathematical theory of communication. *Bell System Technical Journal*, **27**, 379–423.

Shepard, R. N. (1962) The analysis of proximities. Multidimensional scaling with an unknown distance function. *Psychometrika*, **27**, 125–39.

Shwayder, K. (1971) Conversion of limited entry decision tables to computer programs – a proposed modification to Pollack's algorithm. *Communications of the Association of Computer Machinery*, **14**, 69–73.

Shwayder, K. (1974) Extending the information theory approach to converting limited-entry decision tables to computer programs. *Communications of the Association of Computer Machinery*, **17**, 532–37.

Sibson, R. (1970) A model for taxonomy II. *Mathematical Bioscience*, **6**, 405–30.

Sibson, R. (1972) Order invariant methods in data analysis. *Journal of the Royal Statistical Society, Series B*, 311–38.

Simpson, G. G. (1961) *Principles of Animal Taxonomy*. New York: Columbia University Press.

Smirnov, E. S. (1960) Taxonomic analysis of a genus. *Zhurnal Obshenye Biologii*, **21**, 89–103.

Smirnov, E. S. (1969) *Taxonomic Analysis*. Moscow: State University of Moscow.

Sneath, P. H. A. (1968) Vigour and pattern in taxonomy. *Journal of General Microbiology*, **54**, 1–11.

Sneath, P. H. A. (1975) Cladistic representation of reticulate evolution. *Systematic Zoology*, **24**, 360–68

Sneath, P. H. A. (1978a) Identification of microorganisms. In *Essays in Microbiology*, ed. J. R. Norris and M. H. Richmond, pp. 10/1–10/32. London: Wiley.

Sneath, P. H. A. (1978*b*) Classification of microorganisms. In *Essays in Microbiology*, ed. J. R. Norris and M. H. Richmond, pp. 9/1–9/31. London: Wiley.

Sneath, P. H. A. & Sokal, R. R. (1973) *Numerical Taxonomy*. London: Freeman.

Sokal, R. R. & Michener, C. D. (1958) A statistical method for evaluating systematic relationships. *University of Kansas Scientific Bulletin*, **38**, 1409–38.

Sokal, R. R. & Rohlf, F. J. (1962) The comparison of dendrograms by objective methods. *Taxon*, **11**, 33–40.

Sokal, R. R. & Rohlf, F. J. (1970) The intelligent ignoramus, an experiment in numerical taxonomy. *Taxon*, **19**, 305–19.

Sokal, R. R. & Sneath, P. H. A. (1963) *Numerical Taxonomy*. London: Freeman.

Stanier, R. Y., Doudoroff, M. & Adelberg, E. A. (1971) *General Microbiology* (3rd edn). London: Macmillan.

Strauss, D. J. (1975) Model for clustering. *Biometrika*, **62**, 467–75.

Swann, J. M. A. (1970) An examination of some ordination problems by use of simulated vegetational data. *Ecology*, **51**, 89–102.

Temple, J. J. (1968) The low Llandovery (Silurian) brachiopods from Kersley Westmorland. *Paleontology Society Publication No. 521*. London: Paleontology Society.

Thompson, E. A. (1975) *Human Evolutionary Trees*. Cambridge: Cambridge University Press

Ward, J. H. (1963) Hierarchical grouping to optimize an objective function. *Journal of the American Statistical Association*, **58**, 236–44.

Willcox, W. R., Lapage, S. P. & Holmes, B. (1980) A review of numerical methods in bacterial identification. *Antonie van Leeuwenhoek*, **46**, 233–99.

Williams, W. T. & Dale, M. B. (1965) Fundamental problems in numerical taxonomy. In *Advances in Botanical Research*, ed. R. D. Preston, pp. 35–68. London: Academic.

Williams, W. T., Lambert, J. M. & Lance, G. N. (1966) Multivariate methods in plant ecology. V. Similarity analysis and information analysis. *Journal of Ecology*, **54**, 427–45.

Williams, W. T., Lance, G. N., Dale, M. B. & Clifford, H. T. (1971) Controversy concerning the criteria for taxonometric strategies. *Computer Journal*, **14**, 162–5.

Williamson, M. H. (1978) The ordination of incidence data. *Journal of Ecology*, **66**, 911–20.

Wilson, A. C., Carlson, S. S. & White, T. J. (1977) Biochemical evolution. *Annual Review of Biochemistry*, **46**, 573–639.

Woese, C. R., Gibson, J. & Fox, G. E. (1980) Do genealogical patterns in purple photosynthetic bacteria reflect interspecific gene transfer? *Nature*, **283**, 212–14.

Wolfe, J. H. (1970) Pattern clustering by multivariate mixture analysis. *Multivariate Behavioural Research*, **5**, 329–50.

AUTHOR INDEX

SUBJECT INDEX

A CATALOG OF SELECTED
DOVER BOOKS
IN SCIENCE AND MATHEMATICS

A CATALOG OF SELECTED

DOVER BOOKS

IN SCIENCE AND MATHEMATICS

Astronomy

BURNHAM'S CELESTIAL HANDBOOK, Robert Burnham, Jr. Thorough guide to the stars beyond our solar system. Exhaustive treatment. Alphabetical by constellation: Andromeda to Cetus in Vol. 1; Chamaeleon to Orion in Vol. 2; and Pavo to Vulpecula in Vol. 3. Hundreds of illustrations. Index in Vol. 3. 2,000pp. 6⅛ x 9¼.
23567-X, 23568-8, 23673-0 Three-vol. set

THE EXTRATERRESTRIAL LIFE DEBATE, 1750–1900, Michael J. Crowe. First detailed, scholarly study in English of the many ideas that developed from 1750 to 1900 regarding the existence of intelligent extraterrestrial life. Examines ideas of Kant, Herschel, Voltaire, Percival Lowell, many other scientists and thinkers. 16 illustrations. 704pp. 5⅜ x 8½. 40675-X

A HISTORY OF ASTRONOMY, A. Pannekoek. Well-balanced, carefully reasoned study covers such topics as Ptolemaic theory, work of Copernicus, Kepler, Newton, Eddington's work on stars, much more. Illustrated. References. 521pp. 5⅜ x 8½.
65994-1

AMATEUR ASTRONOMER'S HANDBOOK, J. B. Sidgwick. Timeless, comprehensive coverage of telescopes, mirrors, lenses, mountings, telescope drives, micrometers, spectroscopes, more. 189 illustrations. 576pp. 5⅝ x 8¼. (Available in U.S. only.)
24034-7

STARS AND RELATIVITY, Ya. B. Zel'dovich and I. D. Novikov. Vol. 1 of *Relativistic Astrophysics* by famed Russian scientists. General relativity, properties of matter under astrophysical conditions, stars, and stellar systems. Deep physical insights, clear presentation. 1971 edition. References. 544pp. 5⅝ x 8¼. 69424-0

Chemistry

CHEMICAL MAGIC, Leonard A. Ford. Second Edition, Revised by E. Winston Grundmeier. Over 100 unusual stunts demonstrating cold fire, dust explosions, much more. Text explains scientific principles and stresses safety precautions. 128pp. 5⅜ x 8½. 67628-5

THE DEVELOPMENT OF MODERN CHEMISTRY, Aaron J. Ihde. Authoritative history of chemistry from ancient Greek theory to 20th-century innovation. Covers major chemists and their discoveries. 209 illustrations. 14 tables. Bibliographies. Indices. Appendices. 851pp. 5⅜ x 8½. 64235-6

CATALYSIS IN CHEMISTRY AND ENZYMOLOGY, William P. Jencks. Exceptionally clear coverage of mechanisms for catalysis, forces in aqueous solution, carbonyl- and acyl-group reactions, practical kinetics, more. 864pp. 5⅜ x 8½.
65460-5

THE HISTORICAL BACKGROUND OF CHEMISTRY, Henry M. Leicester. Evolution of ideas, not individual biography. Concentrates on formulation of a coherent set of chemical laws. 260pp. 5⅜ x 8½. 61053-5

A SHORT HISTORY OF CHEMISTRY, J. R. Partington. Classic exposition explores origins of chemistry, alchemy, early medical chemistry, nature of atmosphere, theory of valency, laws and structure of atomic theory, much more. 428pp. 5⅜ x 8½. (Available in U.S. only.) 65977-1

GENERAL CHEMISTRY, Linus Pauling. Revised 3rd edition of classic first-year text by Nobel laureate. Atomic and molecular structure, quantum mechanics, statistical mechanics, thermodynamics correlated with descriptive chemistry. Problems. 992pp. 5⅜ x 8½. 65622-5

Engineering

DE RE METALLICA, Georgius Agricola. The famous Hoover translation of greatest treatise on technological chemistry, engineering, geology, mining of early modern times (1556). All 289 original woodcuts. 638pp. 6¾ x 11. 60006-8

FUNDAMENTALS OF ASTRODYNAMICS, Roger Bate et al. Modern approach developed by U.S. Air Force Academy. Designed as a first course. Problems, exercises. Numerous illustrations. 455pp. 5⅜ x 8½. 60061-0

DYNAMICS OF FLUIDS IN POROUS MEDIA, Jacob Bear. For advanced students of ground water hydrology, soil mechanics and physics, drainage and irrigation engineering and more. 335 illustrations. Exercises, with answers. 784pp. 6⅛ x 9¼. 65675-6

ANALYTICAL MECHANICS OF GEARS, Earle Buckingham. Indispensable reference for modern gear manufacture covers conjugate gear-tooth action, gear-tooth profiles of various gears, many other topics. 263 figures. 102 tables. 546pp. 5⅜ x 8½. 65712-4

MECHANICS, J. P. Den Hartog. A classic introductory text or refresher. Hundreds of applications and design problems illuminate fundamentals of trusses, loaded beams and cables, etc. 334 answered problems. 462pp. 5⅜ x 8½. 60754-2

MECHANICAL VIBRATIONS, J. P. Den Hartog. Classic textbook offers lucid explanations and illustrative models, applying theories of vibrations to a variety of practical industrial engineering problems. Numerous figures. 233 problems, solutions. Appendix. Index. Preface. 436pp. 5⅜ x 8½. 64785-4

STRENGTH OF MATERIALS, J. P. Den Hartog. Full, clear treatment of basic material (tension, torsion, bending, etc.) plus advanced material on engineering methods, applications. 350 answered problems. 323pp. 5⅜ x 8½. 60755-0

A HISTORY OF MECHANICS, René Dugas. Monumental study of mechanical principles from antiquity to quantum mechanics. Contributions of ancient Greeks, Galileo, Leonardo, Kepler, Lagrange, many others. 671pp. 5⅜ x 8½. 65632-2

METAL FATIGUE, N. E. Frost, K. J. Marsh, and L. P. Pook. Definitive, clearly written, and well-illustrated volume addresses all aspects of the subject, from the historical development of understanding metal fatigue to vital concepts of the cyclic stress that causes a crack to grow. Includes 7 appendixes. 544pp. 5⅜ x 8½. 40927-9

STATISTICAL MECHANICS: Principles and Applications, Terrell L. Hill. Standard text covers fundamentals of statistical mechanics, applications to fluctuation theory, imperfect gases, distribution functions, more. 448pp. 5⅜ x 8½. 65390-0

THE VARIATIONAL PRINCIPLES OF MECHANICS, Cornelius Lanczos. Graduate level coverage of calculus of variations, equations of motion, relativistic mechanics, more. First inexpensive paperbound edition of classic treatise. Index. Bibliography. 418pp. 5⅜ x 8½. 65067-7

THE VARIOUS AND INGENIOUS MACHINES OF AGOSTINO RAMELLI: A Classic Sixteenth-Century Illustrated Treatise on Technology, Agostino Ramelli. One of the most widely known and copied works on machinery in the 16th century. 194 detailed plates of water pumps, grain mills, cranes, more. 608pp. 9 x 12.
28180-9

ORDINARY DIFFERENTIAL EQUATIONS AND STABILITY THEORY: An Introduction, David A. Sánchez. Brief, modern treatment. Linear equation, stability theory for autonomous and nonautonomous systems, etc. 164pp. 5⅜ x 8¼.
63828-6

ROTARY WING AERODYNAMICS, W. Z. Stepniewski. Clear, concise text covers aerodynamic phenomena of the rotor and offers guidelines for helicopter performance evaluation. Originally prepared for NASA. 537 figures. 640pp. 6⅛ x 9¼.
64647-5

INTRODUCTION TO SPACE DYNAMICS, William Tyrrell Thomson. Comprehensive, classic introduction to space-flight engineering for advanced undergraduate and graduate students. Includes vector algebra, kinematics, transformation of coordinates. Bibliography. Index. 352pp. 5⅜ x 8½. 65113-4

HISTORY OF STRENGTH OF MATERIALS, Stephen P. Timoshenko. Excellent historical survey of the strength of materials with many references to the theories of elasticity and structure. 245 figures. 452pp. 5⅜ x 8½. 61187-6

ANALYTICAL FRACTURE MECHANICS, David J. Unger. Self-contained text supplements standard fracture mechanics texts by focusing on analytical methods for determining crack-tip stress and strain fields. 336pp. 6⅛ x 9¼. 41737-9

Mathematics

HANDBOOK OF MATHEMATICAL FUNCTIONS WITH FORMULAS, GRAPHS, AND MATHEMATICAL TABLES, edited by Milton Abramowitz and Irene A. Stegun. Vast compendium: 29 sets of tables, some to as high as 20 places. 1,046pp. 8 x 10½. 61272-4

FUNCTIONAL ANALYSIS (Second Corrected Edition), George Bachman and Lawrence Narici. Excellent treatment of subject geared toward students with background in linear algebra, advanced calculus, physics and engineering. Text covers introduction to inner-product spaces, normed, metric spaces, and topological spaces; complete orthonormal sets, the Hahn-Banach Theorem and its consequences, and many other related subjects. 1966 ed. 544pp. 6⅛ x 9¼. 40251-7

ASYMPTOTIC EXPANSIONS OF INTEGRALS, Norman Bleistein & Richard A. Handelsman. Best introduction to important field with applications in a variety of scientific disciplines. New preface. Problems. Diagrams. Tables. Bibliography. Index. 448pp. 5⅜ x 8½. 65082-0

FAMOUS PROBLEMS OF GEOMETRY AND HOW TO SOLVE THEM, Benjamin Bold. Squaring the circle, trisecting the angle, duplicating the cube: learn their history, why they are impossible to solve, then solve them yourself. 128pp. 5⅜ x 8½. 24297-8

VECTOR AND TENSOR ANALYSIS WITH APPLICATIONS, A. I. Borisenko and I. E. Tarapov. Concise introduction. Worked-out problems, solutions, exercises. 257pp. 5⅝ x 8¼. 63833-2

THE ABSOLUTE DIFFERENTIAL CALCULUS (CALCULUS OF TENSORS), Tullio Levi-Civita. Great 20th-century mathematician's classic work on material necessary for mathematical grasp of theory of relativity. 452pp. 5⅝ x 8¼. 63401-9

AN INTRODUCTION TO ORDINARY DIFFERENTIAL EQUATIONS, Earl A. Coddington. A thorough and systematic first course in elementary differential equations for undergraduates in mathematics and science, with many exercises and problems (with answers). Index. 304pp. 5⅜ x 8½. 65942-9

FOURIER SERIES AND ORTHOGONAL FUNCTIONS, Harry F. Davis. An incisive text combining theory and practical example to introduce Fourier series, orthogonal functions and applications of the Fourier method to boundary-value problems. 570 exercises. Answers and notes. 416pp. 5⅜ x 8½. 65973-9

COMPUTABILITY AND UNSOLVABILITY, Martin Davis. Classic graduate-level introduction to theory of computability, usually referred to as theory of recurrent functions. New preface and appendix. 288pp. 5⅜ x 8½. 61471-9

ASYMPTOTIC METHODS IN ANALYSIS, N. G. de Bruijn. An inexpensive, comprehensive guide to asymptotic methods–the pioneering work that teaches by explaining worked examples in detail. Index. 224pp. 5⅜ x 8½ 64221-6

ESSAYS ON THE THEORY OF NUMBERS, Richard Dedekind. Two classic essays by great German mathematician: on the theory of irrational numbers; and on transfinite numbers and properties of natural numbers. 115pp. 5⅜ x 8½. 21010-3

APPLIED COMPLEX VARIABLES, John W. Dettman. Step-by-step coverage of fundamentals of analytic function theory–plus lucid exposition of five important applications: Potential Theory; Ordinary Differential Equations; Fourier Transforms; Laplace Transforms; Asymptotic Expansions. 66 figures. Exercises at chapter ends. 512pp. 5⅜ x 8½. 64670-X

INTRODUCTION TO LINEAR ALGEBRA AND DIFFERENTIAL EQUATIONS, John W. Dettman. Excellent text covers complex numbers, determinants, orthonormal bases, Laplace transforms, much more. Exercises with solutions. Undergraduate level. 416pp. 5⅜ x 8½. 65191-6

MATHEMATICAL METHODS IN PHYSICS AND ENGINEERING, John W. Dettman. Algebraically based approach to vectors, mapping, diffraction, other topics in applied math. Also generalized functions, analytic function theory, more. Exercises. 448pp. 5⅝ x 8¼. 65649-7

CALCULUS OF VARIATIONS WITH APPLICATIONS, George M. Ewing. Applications-oriented introduction to variational theory develops insight and promotes understanding of specialized books, research papers. Suitable for advanced undergraduate/graduate students as primary, supplementary text. 352pp. 5⅜ x 8½. 64856-7

COMPLEX VARIABLES, Francis J. Flanigan. Unusual approach, delaying complex algebra till harmonic functions have been analyzed from real variable viewpoint. Includes problems with answers. 364pp. 5⅜ x 8½. 61388-7

AN INTRODUCTION TO THE CALCULUS OF VARIATIONS, Charles Fox. Graduate-level text covers variations of an integral, isoperimetrical problems, least action, special relativity, approximations, more. References. 279pp. 5⅜ x 8½. 65499-0

CATASTROPHE THEORY FOR SCIENTISTS AND ENGINEERS, Robert Gilmore. Advanced-level treatment describes mathematics of theory grounded in the work of Poincaré, R. Thom, other mathematicians. Also important applications to problems in mathematics, physics, chemistry and engineering. 1981 edition. References. 28 tables. 397 black-and-white illustrations. xvii + 666pp. 6⅛ x 9¼. 67539-4

INTRODUCTION TO DIFFERENCE EQUATIONS, Samuel Goldberg. Exceptionally clear exposition of important discipline with applications to sociology, psychology, economics. Many illustrative examples; over 250 problems. 260pp. 5⅜ x 8½. 65084-7

NUMERICAL METHODS FOR SCIENTISTS AND ENGINEERS, Richard Hamming. Classic text stresses frequency approach in coverage of algorithms, polynomial approximation, Fourier approximation, exponential approximation, other topics. Revised and enlarged 2nd edition. 721pp. 5⅜ x 8½. 65241-6

INTRODUCTION TO NUMERICAL ANALYSIS (2nd Edition), F. B. Hildebrand. Classic, fundamental treatment covers computation, approximation, interpolation, numerical differentiation and integration, other topics. 150 new problems. 669pp. 5⅜ x 8½. 65363-3

THE FUNCTIONS OF MATHEMATICAL PHYSICS, Harry Hochstadt. Comprehensive treatment of orthogonal polynomials, hypergeometric functions, Hill's equation, much more. Bibliography. Index. 322pp. 5⅜ x 8½. 65214-9

THREE PEARLS OF NUMBER THEORY, A. Y. Khinchin. Three compelling puzzles require proof of a basic law governing the world of numbers. Challenges concern van der Waerden's theorem, the Landau-Schnirelmann hypothesis and Mann's theorem, and a solution to Waring's problem. Solutions included. 64pp. 5⅜ x 8½. 40026-3

CALCULUS REFRESHER FOR TECHNICAL PEOPLE, A. Albert Klaf. Covers important aspects of integral and differential calculus via 756 questions. 566 problems, most answered. 431pp. 5⅜ x 8½. 20370-0

THE PHILOSOPHY OF MATHEMATICS: An Introductory Essay, Stephan Körner. Surveys the views of Plato, Aristotle, Leibniz & Kant concerning propositions and theories of applied and pure mathematics. Introduction. Two appendices. Index. 198pp. 5⅜ x 8½. 25048-2

INTRODUCTORY REAL ANALYSIS, A.N. Kolmogorov, S. V. Fomin. Translated by Richard A. Silverman. Self-contained, evenly paced introduction to real and functional analysis. Some 350 problems. 403pp. 5⅜ x 8½. 61226-0

APPLIED ANALYSIS, Cornelius Lanczos. Classic work on analysis and design of finite processes for approximating solution of analytical problems. Algebraic equations, matrices, harmonic analysis, quadrature methods, much more. 559pp. 5⅜ x 8½. 65656-X

AN INTRODUCTION TO ALGEBRAIC STRUCTURES, Joseph Landin. Superb self-contained text covers "abstract algebra": sets and numbers, theory of groups, theory of rings, much more. Numerous well-chosen examples, exercises. 247pp. 5⅜ x 8½. 65940-2

SPECIAL FUNCTIONS, N. N. Lebedev. Translated by Richard Silverman. Famous Russian work treating more important special functions, with applications to specific problems of physics and engineering. 38 figures. 308pp. 5⅜ x 8½. 60624-4

QUALITATIVE THEORY OF DIFFERENTIAL EQUATIONS, V. V. Nemytskii and V.V. Stepanov. Classic graduate-level text by two prominent Soviet mathematicians covers classical differential equations as well as topological dynamics and ergodic theory. Bibliographies. 523pp. 5⅜ x 8½. 65954-2

NUMBER THEORY AND ITS HISTORY, Oystein Ore. Unusually clear, accessible introduction covers counting, properties of numbers, prime numbers, much more. Bibliography. 380pp. 5⅜ x 8½. 65620-9

THEORY OF MATRICES, Sam Perlis. Outstanding text covering rank, nonsingularity and inverses in connection with the development of canonical matrices under the relation of equivalence, and without the intervention of determinants. Includes exercises. 237pp. 5⅜ x 8½. 66810-X

INTRODUCTION TO ANALYSIS, Maxwell Rosenlicht. Unusually clear, accessible coverage of set theory, real number system, metric spaces, continuous functions, Riemann integration, multiple integrals, more. Wide range of problems. Undergraduate level. Bibliography. 254pp. 5⅜ x 8½. 65038-3

MODERN NONLINEAR EQUATIONS, Thomas L. Saaty. Emphasizes practical solution of problems; covers seven types of equations. ". . . a welcome contribution to the existing literature...."–*Math Reviews*. 490pp. 5⅜ x 8½. 64232-1

MATRICES AND LINEAR ALGEBRA, Hans Schneider and George Phillip Barker. Basic textbook covers theory of matrices and its applications to systems of linear equations and related topics such as determinants, eigenvalues and differential equations. Numerous exercises. 432pp. 5⅜ x 8½. 66014-1

MATHEMATICS APPLIED TO CONTINUUM MECHANICS, Lee A. Segel. Analyzes models of fluid flow and solid deformation. For upper-level math, science and engineering students. 608pp. 5⅜ x 8½. 65369-2

ELEMENTS OF REAL ANALYSIS, David A. Sprecher. Classic text covers fundamental concepts, real number system, point sets, functions of a real variable, Fourier series, much more. Over 500 exercises. 352pp. 5⅜ x 8½. 65385-4

AN INTRODUCTION TO MATRICES, SETS AND GROUPS FOR SCIENCE STUDENTS, G. Stephenson. Concise, readable text introduces sets, groups, and most importantly, matrices to undergraduate students of physics, chemistry, and engineering. Problems. 164pp. 5⅜ x 8½. 65077-4

SET THEORY AND LOGIC, Robert R. Stoll. Lucid introduction to unified theory of mathematical concepts. Set theory and logic seen as tools for conceptual understanding of real number system. 496pp. 5⅜ x 8¼. 63829-4

TENSOR CALCULUS, J.L. Synge and A. Schild. Widely used introductory text covers spaces and tensors, basic operations in Riemannian space, non-Riemannian spaces, etc. 324pp. 5⅜ x 8¼. 63612-7

ORDINARY DIFFERENTIAL EQUATIONS, Morris Tenenbaum and Harry Pollard. Exhaustive survey of ordinary differential equations for undergraduates in mathematics, engineering, science. Thorough analysis of theorems. Diagrams. Bibliography. Index. 818pp. 5⅜ x 8½. 64940-7

INTEGRAL EQUATIONS, F. G. Tricomi. Authoritative, well-written treatment of extremely useful mathematical tool with wide applications. Volterra Equations, Fredholm Equations, much more. Advanced undergraduate to graduate level. Exercises. Bibliography. 238pp. 5⅜ x 8½. 64828-1

FOURIER SERIES, Georgi P. Tolstov. Translated by Richard A. Silverman. A valuable addition to the literature on the subject, moving clearly from subject to subject and theorem to theorem. 107 problems, answers. 336pp. 5⅜ x 8½. 63317-9

POPULAR LECTURES ON MATHEMATICAL LOGIC, Hao Wang. Noted logician's lucid treatment of historical developments, set theory, model theory, recursion theory and constructivism, proof theory, more. 3 appendixes. Bibliography. 1981 edition. ix + 283pp. 5⅜ x 8½. 67632-3

CALCULUS OF VARIATIONS, Robert Weinstock. Basic introduction covering isoperimetric problems, theory of elasticity, quantum mechanics, electrostatics, etc. Exercises throughout. 326pp. 5⅜ x 8½. 63069-2

THE CONTINUUM: A Critical Examination of the Foundation of Analysis, Hermann Weyl. Classic of 20th-century foundational research deals with the conceptual problem posed by the continuum. 156pp. 5⅜ x 8½. 67982-9

CHALLENGING MATHEMATICAL PROBLEMS WITH ELEMENTARY SOLUTIONS, A. M. Yaglom and I. M. Yaglom. Over 170 challenging problems on probability theory, combinatorial analysis, points and lines, topology, convex polygons, many other topics. Solutions. Total of 445pp. 5⅜ x 8½. Two-vol. set.
Vol. I: 65536-9   Vol. II: 65537-7

A SURVEY OF NUMERICAL MATHEMATICS, David M. Young and Robert Todd Gregory. Broad self-contained coverage of computer-oriented numerical algorithms for solving various types of mathematical problems in linear algebra, ordinary and partial, differential equations, much more. Exercises. Total of 1,248pp. 5⅜ x 8½. Two volumes.
Vol. I: 65691-8   Vol. II: 65692-6

INTRODUCTION TO PARTIAL DIFFERENTIAL EQUATIONS WITH APPLICATIONS, E. C. Zachmanoglou and Dale W. Thoe. Essentials of partial differential equations applied to common problems in engineering and the physical sciences. Problems and answers. 416pp. 5⅜ x 8½. 65251-3

THE THEORY OF GROUPS, Hans J. Zassenhaus. Well-written graduate-level text acquaints reader with group-theoretic methods and demonstrates their usefulness in mathematics. Axioms, the calculus of complexes, homomorphic mapping, p-group theory, more. Many proofs shorter and more transparent than older ones. 276pp. 5⅜ x 8½. 40922-8

DISTRIBUTION THEORY AND TRANSFORM ANALYSIS: An Introduction to Generalized Functions, with Applications, A. H. Zemanian. Provides basics of distribution theory, describes generalized Fourier and Laplace transformations. Numerous problems. 384pp. 5⅜ x 8½. 65479-6

Math–Decision Theory, Statistics, Probability

ELEMENTARY DECISION THEORY, Herman Chernoff and Lincoln E. Moses. Clear introduction to statistics and statistical theory covers data processing, probability and random variables, testing hypotheses, much more. Exercises. 364pp. 5⅜ x 8½. 65218-1

CATALOG OF DOVER BOOKS

STATISTICS MANUAL, Edwin L. Crow et al. Comprehensive, practical collection of classical and modern methods prepared by U.S. Naval Ordnance Test Station. Stress on use. Basics of statistics assumed. 288pp. 5⅜ x 8½.   60599-X

SOME THEORY OF SAMPLING, William Edwards Deming. Analysis of the problems, theory and design of sampling techniques for social scientists, industrial managers and others who find statistics important at work. 61 tables. 90 figures. xvii +602pp. 5⅜ x 8½.   64684-X

STATISTICAL ADJUSTMENT OF DATA, W. Edwards Deming. Introduction to basic concepts of statistics, curve fitting, least squares solution, conditions without parameter, conditions containing parameters. 26 exercises worked out. 271pp. 5⅜ x 8½.   64685-8

LINEAR PROGRAMMING AND ECONOMIC ANALYSIS, Robert Dorfman, Paul A. Samuelson and Robert M. Solow. First comprehensive treatment of linear programming in standard economic analysis. Game theory, modern welfare economics, Leontief input-output, more. 525pp. 5⅜ x 8½.   65491-5

DICTIONARY/OUTLINE OF BASIC STATISTICS, John E. Freund and Frank J. Williams. A clear concise dictionary of over 1,000 statistical terms and an outline of statistical formulas covering probability, nonparametric tests, much more. 208pp. 5⅜ x 8½.   66796-0

PROBABILITY: An Introduction, Samuel Goldberg. Excellent basic text covers set theory, probability theory for finite sample spaces, binomial theorem, much more. 360 problems. Bibliographies. 322pp. 5⅜ x 8½.   65252-1

GAMES AND DECISIONS: Introduction and Critical Survey, R. Duncan Luce and Howard Raiffa. Superb nontechnical introduction to game theory, primarily applied to social sciences. Utility theory, zero-sum games, n-person games, decision-making, much more. Bibliography. 509pp. 5⅜ x 8½.   65943-7

FIFTY CHALLENGING PROBLEMS IN PROBABILITY WITH SOLUTIONS, Frederick Mosteller. Remarkable puzzlers, graded in difficulty, illustrate elementary and advanced aspects of probability. Detailed solutions. 88pp. 5⅜ x 8½.   65355-2

PROBABILITY THEORY: A Concise Course, Y. A. Rozanov. Highly readable, self-contained introduction covers combination of events, dependent events, Bernoulli trials, etc. 148pp. 5⅜ x 8¼.   63544-9

STATISTICAL METHOD FROM THE VIEWPOINT OF QUALITY CONTROL, Walter A. Shewhart. Important text explains regulation of variables, uses of statistical control to achieve quality control in industry, agriculture, other areas. 192pp. 5⅜ x 8½.   65232-7

THE COMPLEAT STRATEGYST: Being a Primer on the Theory of Games of Strategy, J. D. Williams. Highly entertaining classic describes, with many illustrated examples, how to select best strategies in conflict situations. Prefaces. Appendices. 268pp. 5⅜ x 8½.   25101-2

Math–Geometry and Topology

ELEMENTARY CONCEPTS OF TOPOLOGY, Paul Alexandroff. Elegant, intuitive approach to topology from set-theoretic topology to Betti groups; how concepts of topology are useful in math and physics. 25 figures. 57pp. 5⅜ x 8½. 60747-X

COMBINATORIAL TOPOLOGY, P. S. Alexandrov. Clearly written, well-organized, three-part text begins by dealing with certain classic problems without using the formal techniques of homology theory and advances to the central concept, the Betti groups. Numerous detailed examples. 654pp. 5⅜ x 8½. 40179-0

EXPERIMENTS IN TOPOLOGY, Stephen Barr. Classic, lively explanation of one of the byways of mathematics. Klein bottles, Moebius strips, projective planes, map coloring, problem of the Koenigsberg bridges, much more, described with clarity and wit. 43 figures. 210pp. 5⅜ x 8½. 25933-1

CONFORMAL MAPPING ON RIEMANN SURFACES, Harvey Cohn. Lucid, insightful book presents ideal coverage of subject. 334 exercises make book perfect for self-study. 55 figures. 352pp. 5⅝ x 8¼. 64025-6

THE GEOMETRY OF RENÉ DESCARTES, René Descartes. The great work founded analytical geometry. Original French text, Descartes's own diagrams, together with definitive Smith-Latham translation. 244pp. 5⅜ x 8½. 60068-8

THE THIRTEEN BOOKS OF EUCLID'S ELEMENTS, translated with introduction and commentary by Sir Thomas L. Heath. Definitive edition. Textual and linguistic notes, mathematical analysis. 2,500 years of critical commentary. Unabridged. 1,414pp. 5⅜ x 8½. Three-vol. set.
Vol. I: 60088-2 Vol. II: 60089-0 Vol. III: 60090-4

GEOMETRY OF COMPLEX NUMBERS, Hans Schwerdtfeger. Illuminating, widely praised book on analytic geometry of circles, the Moebius transformation, and two-dimensional non-Euclidean geometries. 200pp. 5⅝ x 8¼. 63830-8

DIFFERENTIAL GEOMETRY, Heinrich W. Guggenheimer. Local differential geometry as an application of advanced calculus and linear algebra. Curvature, transformation groups, surfaces, more. Exercises. 62 figures. 378pp. 5⅜ x 8½. 63433-7

CURVATURE AND HOMOLOGY: Enlarged Edition, Samuel I. Goldberg. Revised edition examines topology of differentiable manifolds; curvature, homology of Riemannian manifolds; compact Lie groups; complex manifolds; curvature, homology of Kaehler manifolds. New Preface. Four new appendixes. 416pp. 5⅜ x 8½. 40207-X

TOPOLOGY, John G. Hocking and Gail S. Young. Superb one-year course in classical topology. Topological spaces and functions, point-set topology, much more. Examples and problems. Bibliography. Index. 384pp. 5⅝ x 8¼. 65676-4

LECTURES ON CLASSICAL DIFFERENTIAL GEOMETRY, Second Edition, Dirk J. Struik. Excellent brief introduction covers curves, theory of surfaces, fundamental equations, geometry on a surface, conformal mapping, other topics. Problems. 240pp. 5⅜ x 8½. 65609-8

Math–History of

A SHORT ACCOUNT OF THE HISTORY OF MATHEMATICS, W. W. Rouse Ball. One of clearest, most authoritative surveys from the Egyptians and Phoenicians through 19th-century figures such as Grassmann, Galois, Riemann. Fourth edition. 522pp. 5⅜ x 8½. 20630-0

THE HISTORY OF THE CALCULUS AND ITS CONCEPTUAL DEVELOP-MENT, Carl B. Boyer. Origins in antiquity, medieval contributions, work of Newton, Leibniz, rigorous formulation. Treatment is verbal. 346pp. 5⅜ x 8½. 60509-4

THE HISTORICAL ROOTS OF ELEMENTARY MATHEMATICS, Lucas N. H. Bunt, Phillip S. Jones, and Jack D. Bedient. Fundamental underpinnings of modern arithmetic, algebra, geometry and number systems derived from ancient civiliza-tions. 320pp. 5⅜ x 8½. 25563-8

A HISTORY OF MATHEMATICAL NOTATIONS, Florian Cajori. This classic study notes the first appearance of a mathematical symbol and its origin, the com-petition it encountered, its spread among writers in different countries, its rise to pop-ularity, its eventual decline or ultimate survival. Original 1929 two-volume edition presented here in one volume. xxviii+820pp. 5⅜ x 8½. 67766-4

GAMES, GODS & GAMBLING: A History of Probability and Statistical Ideas, F. N. David. Episodes from the lives of Galileo, Fermat, Pascal, and others illustrate this fascinating account of the roots of mathematics. Features thought-provoking refer-ences to classics, archaeology, biography, poetry. 1962 edition. 304pp. 5⅜ x 8½. (Available in U.S. only.) 40023-9

OF MEN AND NUMBERS: The Story of the Great Mathematicians, Jane Muir. Fascinating accounts of the lives and accomplishments of history's greatest mathe-matical minds–Pythagoras, Descartes, Euler, Pascal, Cantor, many more. Anecdotal, illuminating. 30 diagrams. Bibliography. 256pp. 5⅜ x 8½. 28973-7

HISTORY OF MATHEMATICS, David E. Smith. Nontechnical survey from ancient Greece and Orient to late 19th century; evolution of arithmetic, geometry, trigonometry, calculating devices, algebra, the calculus. 362 illustrations. 1,355pp. 5⅜ x 8½. Two-vol. set. Vol. I: 20429-4 Vol. II: 20430-8

A CONCISE HISTORY OF MATHEMATICS, Dirk J. Struik. The best brief his-tory of mathematics. Stresses origins and covers every major figure from ancient Near East to 19th century. 41 illustrations. 195pp. 5⅜ x 8½. 60255-9

Physics

OPTICAL RESONANCE AND TWO-LEVEL ATOMS, L. Allen and J. H. Eberly. Clear, comprehensive introduction to basic principles behind all quantum optical resonance phenomena. 53 illustrations. Preface. Index. 256pp. 5⅜ x 8½. 65533-4

ULTRASONIC ABSORPTION: An Introduction to the Theory of Sound Absorption and Dispersion in Gases, Liquids and Solids, A. B. Bhatia. Standard reference in the field provides a clear, systematically organized introductory review of fundamental concepts for advanced graduate students, research workers. Numerous diagrams. Bibliography. 440pp. 5⅜ x 8½. 64917-2

QUANTUM THEORY, David Bohm. This advanced undergraduate-level text presents the quantum theory in terms of qualitative and imaginative concepts, followed by specific applications worked out in mathematical detail. Preface. Index. 655pp. 5⅜ x 8½. 65969-0

ATOMIC PHYSICS (8th edition), Max Born. Nobel laureate's lucid treatment of kinetic theory of gases, elementary particles, nuclear atom, wave-corpuscles, atomic structure and spectral lines, much more. Over 40 appendices, bibliography. 495pp. 5⅜ x 8½. 65984-4

AN INTRODUCTION TO HAMILTONIAN OPTICS, H. A. Buchdahl. Detailed account of the Hamiltonian treatment of aberration theory in geometrical optics. Many classes of optical systems defined in terms of the symmetries they possess. Problems with detailed solutions. 1970 edition. xv + 360pp. 5⅜ x 8½. 67597-1

THIRTY YEARS THAT SHOOK PHYSICS: The Story of Quantum Theory, George Gamow. Lucid, accessible introduction to influential theory of energy and matter. Careful explanations of Dirac's anti-particles, Bohr's model of the atom, much more. 12 plates. Numerous drawings. 240pp. 5⅜ x 8½. 24895-X

ELECTRONIC STRUCTURE AND THE PROPERTIES OF SOLIDS: The Physics of the Chemical Bond, Walter A. Harrison. Innovative text offers basic understanding of the electronic structure of covalent and ionic solids, simple metals, transition metals and their compounds. Problems. 1980 edition. 582pp. 6⅛ x 9¼. 66021-4

HYDRODYNAMIC AND HYDROMAGNETIC STABILITY, S. Chandrasekhar. Lucid examination of the Rayleigh-Benard problem; clear coverage of the theory of instabilities causing convection. 704pp. 5⅝ x 8¼. 64071-X

INVESTIGATIONS ON THE THEORY OF THE BROWNIAN MOVEMENT, Albert Einstein. Five papers (1905–8) investigating dynamics of Brownian motion and evolving elementary theory. Notes by R. Fürth. 122pp. 5⅜ x 8½. 60304-0

THE PHYSICS OF WAVES, William C. Elmore and Mark A. Heald. Unique overview of classical wave theory. Acoustics, optics, electromagnetic radiation, more. Ideal as classroom text or for self-study. Problems. 477pp. 5⅜ x 8½. 64926-1

PHYSICAL PRINCIPLES OF THE QUANTUM THEORY, Werner Heisenberg. Nobel Laureate discusses quantum theory, uncertainty, wave mechanics, work of Dirac, Schroedinger, Compton, Wilson, Einstein, etc. 184pp. 5⅜ x 8½. 60113-7

ATOMIC SPECTRA AND ATOMIC STRUCTURE, Gerhard Herzberg. One of best introductions; especially for specialist in other fields. Treatment is physical rather than mathematical. 80 illustrations. 257pp. 5⅜ x 8½. 60115-3

AN INTRODUCTION TO STATISTICAL THERMODYNAMICS, Terrell L. Hill. Excellent basic text offers wide-ranging coverage of quantum statistical mechanics, systems of interacting molecules, quantum statistics, more. 523pp. 5⅜ x 8½. 65242-4

THEORETICAL PHYSICS, Georg Joos, with Ira M. Freeman. Classic overview covers essential math, mechanics, electromagnetic theory, thermodynamics, quantum mechanics, nuclear physics, other topics. First paperback edition. xxiii + 885pp. 5⅜ x 8½. 65227-0

PROBLEMS AND SOLUTIONS IN QUANTUM CHEMISTRY AND PHYSICS, Charles S. Johnson, Jr. and Lee G. Pedersen. Unusually varied problems, detailed solutions in coverage of quantum mechanics, wave mechanics, angular momentum, molecular spectroscopy, more. 280 problems plus 139 supplementary exercises. 430pp. 6½ x 9¼. 65236-X

THEORETICAL SOLID STATE PHYSICS, Vol. 1: Perfect Lattices in Equilibrium; Vol. II: Non-Equilibrium and Disorder, William Jones and Norman H. March. Monumental reference work covers fundamental theory of equilibrium properties of perfect crystalline solids, non-equilibrium properties, defects and disordered systems. Appendices. Problems. Preface. Diagrams. Index. Bibliography. Total of 1,301pp. 5⅜ x 8½. Two volumes. Vol. I: 65015-4   Vol. II: 65016-2

A TREATISE ON ELECTRICITY AND MAGNETISM, James Clerk Maxwell. Important foundation work of modern physics. Brings to final form Maxwell's theory of electromagnetism and rigorously derives his general equations of field theory. 1,084pp. 5⅜ x 8½. Two-vol. set.   Vol. I: 60636-8   Vol. II: 60637-6

OPTICKS, Sir Isaac Newton. Newton's own experiments with spectroscopy, colors, lenses, reflection, refraction, etc., in language the layman can follow. Foreword by Albert Einstein. 532pp. 5⅜ x 8½. 60205-2

THEORY OF ELECTROMAGNETIC WAVE PROPAGATION, Charles Herach Papas. Graduate-level study discusses the Maxwell field equations, radiation from wire antennas, the Doppler effect and more. xiii + 244pp. 5⅜ x 8½. 65678-5

INTRODUCTION TO QUANTUM MECHANICS With Applications to Chemistry, Linus Pauling & E. Bright Wilson, Jr. Classic undergraduate text by Nobel Prize winner applies quantum mechanics to chemical and physical problems. Numerous tables and figures enhance the text. Chapter bibliographies. Appendices. Index. 468pp. 5⅜ x 8½. 64871-0

METHODS OF THERMODYNAMICS, Howard Reiss. Outstanding text focuses on physical technique of thermodynamics, typical problem areas of understanding, and significance and use of thermodynamic potential. 1965 edition. 238pp. 5⅜ x 8½. 69445-3

TENSOR ANALYSIS FOR PHYSICISTS, J. A. Schouten. Concise exposition of the mathematical basis of tensor analysis, integrated with well-chosen physical examples of the theory. Exercises. Index. Bibliography. 289pp. 5⅜ x 8½. 65582-2

RELATIVITY IN ILLUSTRATIONS, Jacob T. Schwartz. Clear nontechnical treatment makes relativity more accessible than ever before. Over 60 drawings illustrate concepts more clearly than text alone. Only high school geometry needed. Bibliography. 128pp. 6⅛ x 9¼. 25965-X

THE ELECTROMAGNETIC FIELD, Albert Shadowitz. Comprehensive undergraduate text covers basics of electric and magnetic fields, builds up to electromagnetic theory. Also related topics, including relativity. Over 900 problems. 768pp. 5⅝ x 8¼. 65660-8

GREAT EXPERIMENTS IN PHYSICS: Firsthand Accounts from Galileo to Einstein, edited by Morris H. Shamos. 25 crucial discoveries: Newton's laws of motion, Chadwick's study of the neutron, Hertz on electromagnetic waves, more. Original accounts clearly annotated. 370pp. 5⅜ x 8½. 25346-5

RELATIVITY, THERMODYNAMICS AND COSMOLOGY, Richard C. Tolman. Landmark study extends thermodynamics to special, general relativity; also applications of relativistic mechanics, thermodynamics to cosmological models. 501pp. 5⅜ x 8½. 65383-8

LIGHT SCATTERING BY SMALL PARTICLES, H. C. van de Hulst. Comprehensive treatment including full range of useful approximation methods for researchers in chemistry, meteorology and astronomy. 44 illustrations. 470pp. 5⅜ x 8½. 64228-3

STATISTICAL PHYSICS, Gregory H. Wannier. Classic text combines thermodynamics, statistical mechanics and kinetic theory in one unified presentation of thermal physics. Problems with solutions. Bibliography. 532pp. 5⅜ x 8½. 65401-X

Paperbound unless otherwise indicated. Available at your book dealer, online at **www.doverpublications.com**, or by writing to Dept. GI, Dover Publications, Inc., 31 East 2nd Street, Mineola, NY 11501. For current price information or for free catalogues (please indicate field of interest), write to Dover Publications or log on to **www.doverpublications.com** and see every Dover book in print. Dover publishes more than 500 books each year on science, elementary and advanced mathematics, biology, music, art, literary history, social sciences, and other areas.